Proceedings in Adaptation, Learning and Optimization

Volume 16

Series Editor

Meng-Hiot Lim, Nanyang Technological University, Singapore, Singapore

The role of adaptation, learning and optimization are becoming increasingly essential and intertwined. The capability of a system to adapt either through modification of its physiological structure or via some revalidation process of internal mechanisms that directly dictate the response or behavior is crucial in many real world applications. Optimization lies at the heart of most machine learning approaches while learning and optimization are two primary means to effect adaptation in various forms. They usually involve computational processes incorporated within the system that trigger parametric updating and knowledge or model enhancement, giving rise to progressive improvement. This book series serves as a channel to consolidate work related to topics linked to adaptation, learning and optimization in systems and structures. Topics covered under this series include:

- complex adaptive systems including evolutionary computation, memetic computing, swarm intelligence, neural networks, fuzzy systems, tabu search, simulated annealing, etc.
- machine learning, data mining & mathematical programming
- hybridization of techniques that span across artificial intelligence and computational intelligence for synergistic alliance of strategies for problem-solving.
- aspects of adaptation in robotics
- agent-based computing
- autonomic/pervasive computing
- dynamic optimization/learning in noisy and uncertain environment
- systemic alliance of stochastic and conventional search techniques
- all aspects of adaptations in man-machine systems.

This book series bridges the dichotomy of modern and conventional mathematical and heuristic/meta-heuristics approaches to bring about effective adaptation, learning and optimization. It propels the maxim that the old and the new can come together and be combined synergistically to scale new heights in problem-solving. To reach such a level, numerous research issues will emerge and researchers will find the book series a convenient medium to track the progresses made.

Indexed by INSPEC, zbMATH.

All books published in the series are submitted for consideration in Web of Science.

Kaj-Mikael Björk
Editor

Proceedings of ELM 2021

Theory, Algorithms and Applications

Springer

Editor
Kaj-Mikael Björk
Graduate School and Research
Arcada University of Applied Sciences
Helsinki, Finland

ISSN 2363-6084 ISSN 2363-6092 (electronic)
Proceedings in Adaptation, Learning and Optimization
ISBN 978-3-031-21677-0 ISBN 978-3-031-21678-7 (eBook)
https://doi.org/10.1007/978-3-031-21678-7

This Springer imprint is published by the registered company Springer Nature Switzerland AG
The registered company address is: Gewerbestrasse 11, 6330 Cham, Switzerland

Organizers

Kaj-Mikael Björk	Graduate School and Research, Arcada University of Applied Sciences, Helsinki, Finland
Professor Jiuwen Cao	Institute of Information and Control, Hangzhou Dianzi University, Xiasha Hangzhou, Zhejiang, 310018, China
Associate Professor Chi-Man Vong	Department of Computer and Information Science, University of Macau, Macau, 999078, China
Dr. Yoan Miche	Bell Labs, Nokia, Espoo, Finland; Department of Information and Computer Science, School of Science, Aalto University, FI-00076, Finland

Contents

Pretrained E-commerce Knowledge Graph Model for Product Classification

Chi-Man Wong[1,2](✉), Chi-Man Vong[2], and Yimin Zhou[3]

[1] Alibaba Group, Hangzhou, China
chiman.wcm@alibaba-inc.com
[2] University of Macao, Macao, China
[3] Chinese Academy of Sciences, Beijing, China

Abstract. Product classification in E-commerce platforms aim to classify a product into correct category, in order to obtain a better guiding service for customer to purchase. Language model is usually used to encode product title into product embedding, and then fed into a classifier for multi-class classification. However, the product title is usually ambiguous and noisy, leading to poor prediction performance. To address this issue, we propose a novel pretrained E-commerce knowledge graph (PEKG) model to learn the representation of product from EKG, and then used it for fine-tuning. We formulate the pretraining on EKG as a multi-view learning problem, where the EKG is divided into four views. From PEKG, user and product representation is proposed to learn via aggregation of its neighbor information, and the semantic meaning from the EKG is learned via translation-based method. We experimentally prove that our proposal significantly outperforms baseline,showing that the PEKG can learn useful representation of product.

Keywords: Pretrained model · Product classification · E-commerce knowledge graph

1 Introduction

E-commerce businesses [12] enable sellers using the Internet and the Web for business transactions of goods, which is affecting much of people's daily life due to its convenience. People can browse and purchase goods from billions of product of all kinds easily, leading to the rapid growing transaction on e-commerce platforms in recent years. Due to the massive scale and abundant variety of products, the organization [2] of the billion items is especially important for guiding customer to purchase, and hence all items should be well categorised. However, the categorization of products is very difficult for seller due to the complexity of the category tree. In real world e-commerce platform, such category tree is usually contained multiple levels which categorise a product from coarse-to-fine level, and each level can be ranged from hundreds to thousands. For example, in Taobao, the number of leaf category is more than 20 thousands. Therefore,

K.-M. Björk (Ed.): ELM 2021, PALO 16, pp. 1–9, 2023.
https://doi.org/10.1007/978-3-031-21678-7_1

it is in need to find a compromising automatic approach to perform accurate product classification for sellers.

Recent approach typically employs natural language processing (NLP) technique [7,21] to learn the embedding representation of product title via language model [3,4], and then such representation is used as input to a neural classifier for multi-class classification. Due to the recent powerful state-of-the-art language model, e.g., BERT [4] (Bidirectional Encoder Representations from Transformers), the classification performance is promising. However, the product title is made up by sellers based on their understanding of the product and the target customers. Such title may not reflect well of the product's category, and most information of the title is usually used to attract customer's attention to click and purchase. For example, a title may be "Creative retro Ceramic Cup Mug simple coffee cup with lid spoon personality frosted couple Japanese drinking cup", which is confusing because there are multiple product types such as "cup", "coffee", and "spoon". Therefore, the information of title may be too noisy to learn the real classification of the product, and more information should be considered to provide a complete picture of the product.

Knowledge Graph (KGs) [19] has been used to organise facts as triples, such as *(iPad,colorIs,red)* in E-commerce platform (e.g, Alibaba or Amazon's Knowledge Graph). The advantage is that multiple sources can be easily fused into the same KG. For example, the user behavior, the attributes of product, the categorization of product and the relation between products can be easily integrated into one KG namely e-commerce knowledge graph (EKG). Such E-commerce knowledge graph contains abundant information of product from multiple views. In Alibaba, the e-commerce KG contains 70+ billion triples. Therefore, in this work, we propose using a pretrained E-commerce knowledge graph (PEKG) model to learn a complete information of product as product embedding, and then together fuse with the title embedding from BERT for product classification.

In summary, contributions of this work are as follows:

- we propose a pretrain and fine-tune framework for product classification, which integrate multiple sources of product-related information;
- a pretrained e-commerce knowledge graph model is proposed to learn the representation of product from the e-commerce knowledge graph, which can overcome the noisy and confusing title-based product classification;
- We test PEKG on our billion-scale e-commerce knowledge graph and test it on product classification, showing that PEKG successfully improves the performance.

2 Related Work

2.1 Pretrained Language Model

In recent years, more and more researchers pay attention to the development of the pretrained language models [3,4], because the pretrained language models have obtained an excellent performance in many NLP tasks. The workflow of the learning process is that a pretrained model is firstly learned on a massive and easily obtainable corpus, and then encode the input data from downstream task into a useful representation; and finally the learned representation are fed into a neural classifier for fine-tuning. The representative pretrained language model are BERT [4] and GPT-3 [3]. Currently, the hybrid approach of knowledge graph and pretrained language model can obtain even better performance [6,9,17,23]. For examples, K-Bert [9] combines the knowledge triples from KG with the sentence input, and employs visible matrix to control the data fusion of the sentence and the knowledge; ERNIE [23] integrates the informative facts into the model via a knowledge module in order to represent the heterogeneous information of tokens and entities into same feature space. K-Adapter [16] employs a fixed-parameter pretrained model (e.g., Roberta), and defines factual and linguistic adapter to learn individually for each task; LIBERT [6] imposes an additional language constraint task to integrates knowledge; KEPLER [17] links the entity with the sentence, and respectively encode the sentence into embedding with BERT, and the entity into entity embedding with transE [1]. Based on BERT or other pre-trained language models, these methods combine the language model with the knowledge, while our pretrained model is learnt only on the triples from the e-commerce knowledge graph.

2.2 Knowledge Graph Embedding

The method of knowledge graph embedding (KGE) [1,5,8,14,18,20] has been successfully applied into many knowledge graph tasks. The famous one is TransE [1], introducing a translation-based method to learn the representation of entities and relations into a continuous vector space. Such Translation-based methods employ a scoring function $f(h,r,t)$ to measure the plausibility of a fact (h,r,t). Since then, more other methods are proposed, such as TransH [18], TransR [8], TransD [5] and DistMult [20]. TransH [18] learns multi-relation by decomposing h, r and t into two parts and projecting them on the hyperplane; TransR [8] projects entities and relations into entity space and multiple relation spaces by projection matrix respectively; TransD [5] represents entity and relation with semantic vectors and projection vectors to model multi-entity spaces and multi-relation spaces; TransEdge [14] learns edge-centric embedding via relation contextualisation; DistMult [20] composes the relation to captures relational semantics. Another famous methods are GCN-based method [10,11,13,15,22]. GCN-based methods learn the representation of entity and relation via the aggregation of neighbor information. For example, GCN [13] introduces a first-order approximation of ChebNet to perform graph convolution; AGCN [10] introduce a distance metric learning to learn graph structure information; DGCN [11] employs normalized adjacency and positive pointwise mutual information (PPMI) matrix to perform dual graph convolution to encodes both local and global structure information. GAT [15] learn the importance of neighbor information with attention mechanism via a masked self-attention layer; Gaan [22] employs sub0convolution network the weights each attention head differently.

We combine transE and GCN based methods in order to learn both structure and semantic information.

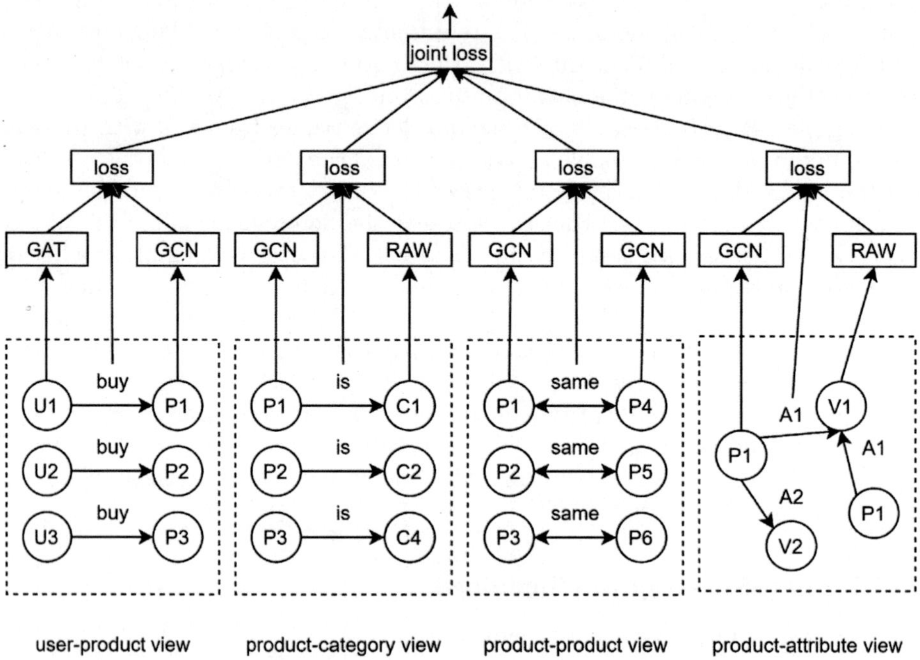

Fig. 1. The architecture of the proposed pretrained model.

3 Proposed Method

Our method contains two new parts: 1) constructing and pre-training a EKG to encode structure and semantic information; 2) fine-tuning a BERT based on normal title embedding and features from pre-trained e-commerce knowledge graph representations.

3.1 Constructing and Pre-training of EKG

E-commerce Knowledge Graph Construction We make knowledge graph as a way to encode diverse information related to product and we construct the EKG from real-world triples.

There are four kinds of information contained in EKG:

- User attributes: each user has a list of attributes, such as gender, etc., which show the user preference. Each user attribute in EKG has the form of (user, has, user attribute).

- Product attributes: our platform contains abundant information of products like categories and attributes such as color, brand, etc. For categories, a triplet in the form of (product, belong to, category), while (product, has, value) for product attributes are built. This product information provides a fruitful information to the product.
- Same Product record: there are multiple of sellers have the same products on their shops, but usually with different attributes and description. Here we has the form of (product, same, product) are built.
- User behavior: each user has multiple interactions with the product, which for example has the form (user, buy, product).

E-commerce Knowledge Graph Pre-training We have formulated the learning of E-commerce knowledge graph as a multi-view learning problem, where each view are learnt with different task. For all views, the embedding of user, product and their attributes are learnt differently.

User Representation. In e-commerce platform, there are certain number of user registered everday, so that using lookup table embedding for user embedding is not feasible. Therefore, in our pretrained, we employ the GAT method to encode the user's attributes as user representation, because some attributes of user such as gender is more important to the user preference. Suppose user u_i has multiple attributes a_j (i.e., $u = [a_1, a_2, a_3, \ldots, a_j]$), the user representation R_u is given by

$$R_u = \sigma(\alpha_{ij}\mathbf{Wu}) \tag{1}$$

where α_{ij} is the attention score on the attributes, $\mathbf{W}$ is the weight matrix to be learnt, and σ is the sigma function.

Product Representation. Same as user representation, many new products are published by seller everday, so that lookup table embedding method is not applicable too. Since the product attribute can reflect the product equally, we employ GCN method to encode the product's attributes as product representation. Suppose product p_i has multiple attributes a_j (i.e., $p = [a_1, a_2, a_3, \ldots, a_j]$), the product representation R_p is given by

$$R_p = \sigma(\mathbf{Wp}) \tag{2}$$

where $\mathbf{W}$ is the weight matrix to be learnt, and σ is the sigma function.

Model Pretraining. Since the number of attributes a_j and relation r_j tend to be stable, we employ lookup embedding table for them. As shown in Figure. 1, After we have the representation of user, product, and their attributes and relation, there are four tasks to model different information of view. Gathering all views can provide more complete picture of product. For each task/view, transE is employed to learn the semantic information for each view. For a triplet (e_i, r, e_j), the definition of the score function is given by

$$f(e_i, r, e_j) = ||\mathbf{e}_i + \mathbf{r} - \mathbf{e}_j|| \tag{3}$$

where $\mathbf{e}_i, \mathbf{r}$ and $\mathbf{e}_j$ are embeddings of e_i, r and e_j respectively. The embedding of $\mathbf{e}_i$ or $\mathbf{e}_j$ are equal to R_u for user and R_p for product, and the lookup embedding for the remaining entities and relations. Then, using this score function, a higher score will be given to positive triplets while lower score given to negative ones. Finally, the training of the model is to minimize a margin-based ranking loss, which is

$$L = \sum_{(e_i, r, e_j) \in T} [f(e_i, r, e_j) + \gamma - f(e_i', r', e_j')]_+ \tag{4}$$

where (e_i', r', e_j') is the negative sample of (e_i, r, e_j) by randomly replacing e_i or e_j with $e \in EKG$. The function $[x]_+ = 0$ if $x < 0$, otherwise $[x]_+ = x$.

Fine-tuning for Product Classification After the construction and pre-training of EKG, the trained entity embeddings are used to fine-tune for product classification. In product classification, product embedding from EKG are used to knowledge-based enhancement of the representation. Given any classifier y and input features A, and the knowledge embeddingK_P, then the enhancement of the input can be given by $[A; K_P]$. Finally the output can be given by

$$y = f_{output}([A; K_P]) \tag{5}$$

where K_P usually remains fixed for the fine-tuning. From this approach, the knowledge embedding can be easily employed in any downstream models for end-to-end training.

4 Experiments

4.1 Datasets & Implementation

PEKG is pretrained and fine-tuned on a private dataset from scratch, which has 60B triples. We sampled 18 categories from different levels to test the performance of our model. As shown in Table. 1, each level has 6 categories for evaluation.We have employed BERT as our base model because it is the state-of-art model on NLP tasks. The parameters of BERT is with hidden size = 768; number of layer = 3; and the hidden layer activation = gelu is employed. For the pretraining of PEKG, we have employed a distribute training with number of ps = 1 and number of workers = 400; and the intial learning rate is set as 0.0001. After the pretraining of PEKG is finished, the embedding for product is obtained as input to BERT for fine-tuning as $BERT_{PEKG}$. The parameter of $BERT_{PEKG}$ is the same as the base model.

4.2 Results

We compare our methods with 18 datasets from coarse to fine levels. From these datasets, we calculate metrics Hit@k(k= 1,5,10) to describe the performance of the model, where we employed the rank of the correct label as its prediction rank

Table 1. The number of samples for each category.

Level	Category	Samples
Coarse	3C Digital Accessories (3C)	515447
	Electric Vehicles/Accessories (EA)	423444
	Video Games (VG)	1072204
	Textile Fabrics (TF)	1465511
	Mechanical Equipment (ME)	462014
	Festival Supplies/Gifts (GF)	931982
Medium	MP3/MP4 Accessories (MA)	671509
	Electric Vehicle Accessories (EA)	391425
	Handheld Accessories (HA)	544371
	Textile Accessories (TA)	903113
	Construction Machinery (CM)	322619
	Creative Gifts (CG)	318817
Fine	Charger (C)	951671
	Electric Vehicle Battery (EB)	614831
	Key (K)	279518
	Webbing (W)	950001
	Excavating Machinery (EM)	510364
	Conch/Shell (CS)	680103

for that category and Hit@k is the percentage of test samples with prediction rank within k. As shown in Table. 2, The proposed $BERT_{PEKG}$ outperforms BERT for all datasets. Specifically, $BERT_{PEKG}$ has an average improvement of 8.05% for hit@1, 7.13% for hit@5, and 4.18% for hit@10, which shows that PEKG can really learn a useful representation of product from E-commerce knowledge graph. Beside, the performance of $BERT_{PEKG}$ has more improvement on the lower k. It is because some product titles are usually confusing, leading BERT performing worse for Hit@1, but better performance for Hit@5 and Hit@10. While in PEKG, the product attribute is learned so that more information can be captured in the model.

5 Conclusion

In this paper, we propose a two-step product classification model. We first propose a pretrained model PEKG based on a E-commerce knowledge graph. Then we propose a fine-tuning appraoch to inject the knowledge embedding to BERT. Experimental results show that our model obtains substantial performance on 18 datasets compared to baselines.

Table 2. The testing performance (%) of $BERT_{PEKG}$ and BERT in 18 categories.

Level	Dataset	BERT			$BERT_{PEKG}$		
		Hit@1	Hit@5	Hit@10	Hit@1	Hit@5	Hit@10
Coarse	3C	67.39	93.48	97.83	73.91	97.83	100.00
	EA	78.57	92.43	96.43	88.37	93.02	97.67
	VG	65.85	90.24	95.12	73.17	92.68	97.56
	TF	66.61	84.85	87.88	77.08	89.58	95.83
	ME	58.57	85.71	92.86	62.79	88.37	95.35
	GF	69.09	90.94	96.56	75.76	93.94	96.97
Medium	MA	43.48	76.09	91.30	50.00	91.30	95.65
	EA	64.29	89.29	96.43	74.42	93.02	95.35
	HA	51.22	82.93	92.18	60.98	92.68	92.68
	TA	64.55	84.85	87.88	72.92	89.58	91.67
	CM	68.57	88.86	92.86	72.09	88.37	93.02
	CG	69.09	85.15	89.88	72.73	87.88	90.94
Fine	C	57.30	69.57	80.43	63.04	89.13	91.30
	EB	78.57	89.29	96.43	79.07	92.35	95.35
	K	43.90	75.61	82.93	58.54	85.37	90.24
	W	42.42	72.73	75.76	58.33	83.33	85.42
	EM	53.57	57.14	71.43	65.12	86.05	88.37
	CS	60.06	69.70	75.70	69.70	72.73	81.82
	Average	61.28	82.16	88.88	69.33	89.29	93.07

References

1. Bordes, A., Usunier, N., García-Durán, A., Weston, J., Yakhnenko, O.: Translating embeddings for modeling multi-relational data. In: NIPS. pp. 2787–2795 (2013)
2. Brisoux, J.E., Cheron, E.J.: Brand categorization and product involvement. ACR North Am. Adv. (1990)
3. Brown, T.B., Mann, B., Ryder, N., Subbiah, M., Kaplan, J., Dhariwal, P., Neelakantan, A., Shyam, P., Sastry, G., Askell, A., et al.: Language models are few-shot learners. arXiv preprint arXiv:2005.14165 (2020)
4. Devlin, J., Chang, M., Lee, K., Toutanova, K.: BERT: pre-training of deep bidirectional transformers for language understanding. In: NAACL-HLT (1). pp. 4171–4186. Association for Computational Linguistics (2019)
5. Ji, G., He, S., Xu, L., Liu, K., Zhao, J.: Knowledge graph embedding via dynamic mapping matrix. In: Proceedings of the 53rd Annual Meeting of the Association for Computational Linguistics and the 7th International Joint Conference on Natural Language Processing of the Asian Federation of Natural Language Processing, ACL 2015, July 26-31, 2015, Beijing, China, Volume 1: Long Papers. pp. 687–696 (2015)

6. Lauscher, A., Vulić, I., Ponti, E.M., Korhonen, A., Glavaš, G.: Specializing unsupervised pretraining models for word-level semantic similarity. arXiv preprint arXiv:1909.02339 (2019)
7. Leukel, J., Schmitz, V., Dorloff, F.D.: A modeling approach for product classification systems. In: Proceedings of 13th International Workshop on Database and Expert Systems Applications, pp. 868–874. IEEE (2002)
8. Lin, Y., Liu, Z., Sun, M., Liu, Y., Zhu, X.: Learning entity and relation embeddings for knowledge graph completion. In: AAAI, pp. 2181–2187. AAAI Press (2015)
9. Liu, W., Zhou, P., Zhao, Z., Wang, Z., Ju, Q., Deng, H., Wang, P.: K-bert: Enabling language representation with knowledge graph. arXiv:1909.07606 (2019)
10. Park, C., Park, J., Park, S.: Agcn: attention-based graph convolutional networks for drug-drug interaction extraction. Expert Syst. Appl. **159**, 113538 (2020)
11. Qiu, Z., Qiu, K., Fu, J., Fu, D.: Dgcn: dynamic graph convolutional network for efficient multi-person pose estimation. In: Proceedings of the AAAI Conference on Artificial Intelligence. vol. 34, pp. 11924–11931 (2020)
12. Reynolds, J.: Ecommerce: a critical review. Int. J. Retail Distrib. Manage. (2000)
13. Schlichtkrull, M., Kipf, T.N., Bloem, P., Van Den Berg, R., Titov, I., Welling, M.: Modeling relational data with graph convolutional networks. In: European Semantic Web Conference. pp. 593–607. Springer (2018)
14. Sun, Z., Huang, J., Hu, W., Chen, M., Guo, L., Qu, Y.: Transedge: Translating relation-contextualized embeddings for knowledge graphs. In: International Semantic Web Conference. pp. 612–629. Springer (2019)
15. Veličković, P., Cucurull, G., Casanova, A., Romero, A., Lio, P., Bengio, Y.: Graph attention networks. arXiv preprint arXiv:1710.10903 (2017)
16. Wang, R., Tang, D., Duan, N., Wei, Z., Huang, X., Cao, C., Jiang, D., Zhou, M., et al.: K-adapter: infusing knowledge into pre-trained models with adapters. arXiv preprint arXiv:2002.01808 (2020)
17. Wang, X., Gao, T., Zhu, Z., Liu, Z., Li, J., Tang, J.: Kepler: a unified model for knowledge embedding and pre-trained language representation. arXiv:1911.06136 (2019)
18. Wang, Z., Zhang, J., Feng, J., Chen, Z.: Knowledge graph embedding by translating on hyperplanes. In: AAAI, pp. 1112–1119. AAAI Press (2014)
19. Xu, D., Ruan, C., Korpeoglu, E., Kumar, S., Achan, K.: Product knowledge graph embedding for e-commerce. In: Proceedings of the 13th International Conference on Web Search and Data Mining. pp. 672–680 (2020)
20. Yang, B., Yih, W., He, X., Gao, J., Deng, L.: Embedding entities and relations for learning and inference in knowledge bases. In: 3rd International Conference on Learning Representations, ICLR 2015, San Diego, CA, USA, May 7-9, 2015, Conference Track Proceedings (2015)
21. Zahavy, T., Magnani, A., Krishnan, A., Mannor, S.: Is a picture worth a thousand words? a deep multi-modal fusion architecture for product classification in e-commerce. arXiv preprint arXiv:1611.09534 (2016)
22. Zhang, J., Shi, X., Xie, J., Ma, H., King, I., Yeung, D.Y.: Gaan: Gated attention networks for learning on large and spatiotemporal graphs. arXiv preprint arXiv:1803.07294 (2018)
23. Zhang, Z., Han, X., Liu, Z., Jiang, X., Sun, M., Liu, Q.: Ernie: Enhanced language representation with informative entities. arXiv preprint arXiv:1905.07129 (2019)

A Novel Methodology for Object Detection in Highly Cluttered Images

Kallin Carolus Khan[1](✉), Edward Ratner[1], Christopher Douglas[1], and Amaury Lendasse[2,3]

[1] Edammo Inc, Iowa City, USA
kallinkhan@gmail.com
[2] Department of ILT, University of Houston, Houston, USA
[3] Arcada University of Applied Sciences, Helsinki, Finland

Abstract. Over the past several decades camera technology has exploded in terms of higher quality, decreased cost, and increased availability. This means that there are significantly more images and videos being collected, which has developed an increased need for machine learning algorithms and methodologies that can extract and leverage relevant information from them. While much work has been done, current algorithms fall short on being able to both locate and classify objects in highly cluttered images, where it is difficult for even a human to identify objects. In this paper, we create and test a novel methodology to perform object detection in highly cluttered images by utilizing the partial least squares algorithm for dimensionality reduction, transfer learning for feature extraction, a newly developed object localization technique, and an ensemble of Extreme Learning Machines for classification. This methodology outperforms the current state of the art, Google's AutoML Vision.

Keywords: Object detection · Image analysis · Localization

1 Introduction

Machine learning algorithms are becoming more and more successful at being able to process an image, locate relevant objects, and label said objects accordingly. However there has been limited development of algorithms that can perform the mentioned series of operations on highly cluttered images. A highly cluttered image is defined as an image where there is more irrelevant information than relevant information. Cluttered images also have the property of being extremely difficult for even humans to identify objects among the clutter. When it comes to the object localization and classification in our project, this irrelevant information comes in the form of irrelevant objects. The irrelevant objects in our highly cluttered images are often larger than and overlap onto the objects that we wish to locate and classify with our algorithm.

Much of the development of algorithms to detect objects in images has been done on objects easily discernible to the human eye, such as cars, animals, and

K.-M. Björk (Ed.): ELM 2021, PALO 16, pp. 10–23, 2023.
https://doi.org/10.1007/978-3-031-21678-7_2

human faces [1,2]. These images may have cluttered backgrounds, but the objects themselves are recognizable and easy to identify. Most machine learning algorithms for object localization and classification in images try to achieve the goal of matching human performance. Due to the difficulty of even humans to recognize objects in highly cluttered images, in this paper we develop an algorithm to perform object localization and classification better than a person could. Here, we propose a novel methodology for object detection in highly cluttered images. Our method is novel in two ways. First, we developed a new algorithm for the localization of objects within an image. And secondly, as stated above, this methodology performs better than a human can, whereas in other experiments the goal is only to match human performance.

Others have used k-Nearest Neighbors, Randomized Neural Networks, and Single-Layer Feed-forward Networks, however in our research we did not come across images nearly as cluttered as ours [3–5]. More often than not, the objects of interest in these other studies took up the majority of the image and were easily identifiable. Our methodology utilized ELMs, and more specifically a One versus All ensemble of ELMs. In Sect. 2, we introduce the concept and mechanics of Extreme Learning Machines. Following, in Sect. 3, we go into detail about the implementation of the traditional ELM ensembling technique, and, in Sect. 4, how we modify this method to the One versus All approach. In Sect. 5, Experimental Data and Settings, we go into the specifics of the problem at hand, namely locating and classifying relevant recyclable objects in images of a moving conveyor belt. The method by which we preprocess and extract features from these images is described in Sect. 6. This is followed by the description of the implementation of our novel algorithm for object localization in Sects. 7 and 8. Lastly, the results of our experiment are summarized and we draw conclusions about their implications in Sects. 9 and 10, respectively.

2 Extreme Learning Machine

The Extreme Learning Machine [6,7] is introduced as a generalized Single-Layer Feed-forward Network (SLFN) [6,8]. This type of Network is capable of solving classification [9], regression [10–14] and clustering problems. According to Huang et al. in [7], ELM has good generalized performance in most cases and the learning speed is thousands of times faster than conventional neural networks [6,8], an advantage that allows the parameters of an ELM to be modified, tested, and tuned much more quickly than other algorithms.

ELM belongs to the family of Randomized Neural Networks (RNNs). Unlike traditional neural networks and learning algorithms, the ELM algorithm shows that hidden nodes can be randomly generated. Thus, the weights from the first layer can be independent from the training data. Because there is no dependence between the input and output weights, ELM has a non-iterative linear ordinary least square solution for the output weights, unlike the conventional Back-propagation training procedure [15]. On top of the distinct properties of ELM, Huang et al. in [8,16] stated that ELM has the universal approximation capability, indicating that ELM can universally approximate any continuous target functions in any compact subset X of the Euclidean space $\mathbb{R}^n$ [17].

3 Traditional ELM Ensembling Technique

As described above in Sect. 2, there is randomization of the weights of the input layer of an ELM. This creates the issue of predictive stability for the algorithm as some ELMs inherently perform worse than others. Therefore in the practiced use of ELMs, a traditional ensemble is utilized. This traditional ensemble is composed of numerous ELMs, and its training has two parts, which requires a split of the training set into two subsets: training and validation.

Starting with the training subset, we normalize the data by using the mean and standard deviations of each feature, then we compute the weights of the output layer for each ELM in the ensemble by the process described in Section II. Once we finish computing these output layer weights, the first part of the training step is complete and we move on to validation. In the second step of training, we utilize each individual model to make predictions on the second subset of our training data, the validation set. Since the input weights of an ELM are randomized, the models within the ensemble have varying levels of performance. By training on a subset of the training data and predicting on the validation set, we evaluate the models that perform best on the data and give them more weight in the final decision. To execute this second step, we first normalize the validation set using the mean and standard deviation of the training set. Then, each model in the ensemble predicts on each sample and is assigned a weight. The weight of each model within the traditional ELM ensemble is equal to its accuracy of classifying correctly the samples in the validation set.

After the models are trained and weighted, the traditional ELM ensemble can make predictions on the testing set. Each model classifies each sample, just as in the training set, however when it comes to choosing a final classification for the sample, the ensemble chooses the class that has the highest "weight sum" as so:

$$\boldsymbol{class} = \arg\max\,(s_1, s_2, \ldots, s_i). \tag{1}$$

where, i equals the number of possible classes, and $s_1, s_2, \ldots, s_i$ represent the weight sums of classes $1, 2, \ldots, i$. The weight sum is calculated by summing the weights of all the models that predicted a sample as a certain class.

This weighting strategy mitigates the issue of predictive stability of an ELM caused by the model's random input layer. By utilizing numerous ELMs, in some cases several hundred, then weighting each model based on its performance predicting the validation set with superior performing ELMs weighted more heavily, and finally using these weights to make an ensembled final classification as described above, it becomes possible to observe consistent and accurate results despite the randomly created input layer of the ELM models. Additionally, since every ELM is weighted based on its accuracy at classifying the validation set, we can assign a minimum threshold accuracy for giving a model any weight at all. This, in turn, would give ELMs that are poor at predicting a certain data set no say in the ensemble while at the same time giving relatively more say to the models whose weight is above the threshold.

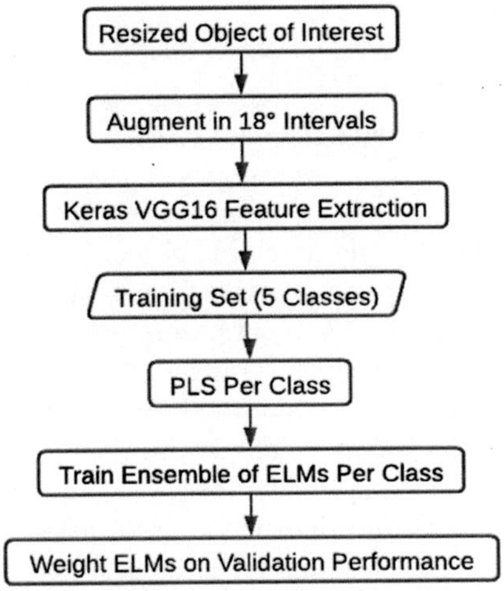

Fig. 1. Data Preprocessing and OvA Training

4 One Versus All ELM Ensembling Technique

The One versus All (OvA) ELM Ensembling Technique is similar to the Traditional ELM Ensembling Technique, with an added step. The OvA method requires an ensemble for every class within the data. Each sub-ensemble has nearly identical training data with the exception of the class of each sample, which changes depending on the sub-ensemble being trained. No matter how many classes are in the training set, only two classes are recognized in a sub-ensemble of an OvA ensemble. Each sub-ensemble has a different "focus class" from the training set, and the sub-ensemble's goal is to discern if a sample belongs to the focus class or not. The focus class is the "one" in each sub-ensemble in the OvA method, such that each class in the data is the corresponding focus class for a single sub-ensemble. Each sample in the training set is specified as class 0 when training the sub-ensemble with the associated focus class and as class 1 for all other sub-ensembles. Once each sample's class is correctly updated for a particular sub-ensemble, the training process is the same as that of the traditional ELM Ensemble described in the previous section. The data is split into training and validation sets, normalized, and then the output layer weights of each ELM is computed. Then, each ELM is given a weight based on its accuracy of prediction on the validation set. This process is repeated for each sub-ensemble, changing the samples' classes based on the focus class of the sub-ensemble being trained. Once all of the sub-ensembles are trained, the OvA Ensemble is ready to predict. The predicting methodology for each sub-ensemble is identical to that of the Traditional ELM Ensemble as described in the previous section, however instead of returning a class using (1), each sub-ensemble returns the weight sum of class 0. This weight sum represents the predicted probability that a test sample belongs to the focus class of a sub-ensemble.

5 Experimental Data and Settings

The data set used in this experiment is a proprietary collection of images of a moving conveyor belt of trash and recyclable objects. Though there are numerous types of objects on the conveyor belt, the four groups that are of interest are (1) aluminum and tin cans, (2) colored high-density polyethylene (HDPE), (3) other HDPE, and (4) polyethylene terephthalate (PETE) plastic objects. Our training set consists of 3,800 objects pulled from conveyor belt images. Each of the four groups has 500 objects in the training set, and the remaining 1,800 consists of parts of the conveyor belt images without objects of interest. These are labeled as "other" so that our algorithm could distinguish between objects of interest and non-interest. The testing set is composed of 100 successive images of the moving conveyor, which contain in total 89 objects of interest to identify. In order to evaluate the performance of our novel method, we trained and tested Google's deep-learning based AutoML Vision Object Detection [18] with the same data and compared the results.

Since there are 5 classes including the other class and we utilize a one versus all ensembling technique, our algorithm has 5 ensembles of ELMs. Each sub-ensemble is composed of 600 ELMs that had varying alpha L_2 normalization, number of neurons, and input weight parameters that make each of them unique. Thus, in total 3000 ELMs are trained and weighted based on their performance in validation, a process that is described in Sections III and IV. Based on this weighting, only the 50 models in each OvA sub-ensemble that predicted most accurately in validation are used in the testing stage.

6 Image Preprocessing and Feature Extraction

The first step in our experiment is to preprocess our images and perform feature extraction on them. We began with a data set of pre-classified images, where the objects of interest are contained in a bounding box. We then pull these objects of interest and resize them to 224×224, which is the default input size for our feature extraction models. The training data set is then split into training and validation. Once split, augmentation of the data in the training set is performed. Augmentation is done by rotating each object in 18°C intervals (i.e. 18, 36, 54, ..., 360), saving each version, taking the mirror image of the object and repeating the rotation, again saving each version.

After augmentation, we leverage transfer learning for feature extraction which is performed by using the Keras VGG16 convolutional neural network.

We chose the VGG16 feature extractor because of previously reported benefits [19,20]. We chose to use an existing feature extraction method and not develop our own because it is not the primary focus of our research and VGG16 features performed well over a wide range of images. Next, we delete features in the feature set where the standard deviation is near 0 (less than 1e-5). This is done because, with no variability, these features provide almost no information for our models and cause predictive instability.

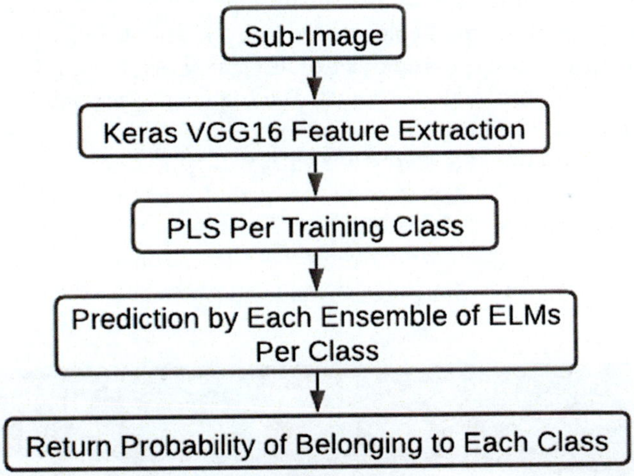

Fig. 2. OvA Scoring Flowchart

Finally, we utilize the partial least squares (PLS) algorithm for dimensionality reduction. PLS is a statistical technique that achieves supervised dimensionality reduction by computing a relationship between observed and latent variables and can be used in both regression and classification contexts [21,22]. This use of PLS was necessary for the experiment because our data has 50,176 features, a significantly larger figure than the number of samples we are training on. Given that PLS utilizes both the features of a sample as well as its classification, we perform the algorithm within the OvA ensemble a total of 5 times, once for every sub-ensemble. Therefore, we are able to reduce the number of features for each specific class independently. The 50,176 features for aluminum and tin cans, colored HDPE, other HDPE, PETE, and non-interest objects are reduced to feature set sizes of 21, 35, 35, 19, and 19, respectively, using the PLS algorithm. The features set sizes are tuned to these values based on their performance in validation. Once this preprocessing is complete, we train the OvA ensemble. This process is illustrated in Fig. 1.

Nearly the entire process above is then repeated, with two changes. First, after pulling the objects of interest from the pre-classified images, the objects are resized to 32×32 instead of 224×224. The second change is that PLS is not used and the dimensionality is not reduced for these lower resolution images.

The reason for processing the objects of interest as both 224×224 and 32×32 is explained further in the following section, but in short this is done to speed up analysis in testing. 224×224 images take much longer to process than 32×32, so, in testing, our algorithm first calculates the probability score of containing an object of interest for all 32×32 sub-images in a conveyor belt image, and those most likely containing an object of interest are further analyzed in the higher resolution of 224×224.

Before we took the results from our 32×32 image pipeline into the 224×224, we did some visual analysis of the results we were getting via multi-color heatmaps. We generated small color-coded images where each pixel of the image represented the probability of an object of interest. Each color represents a different class, and the higher the color value, the higher the probability of an object being there. This method allowed us not only to see what our system was predicting, but also to refine our algorithm with real time feedback in order to have the best prediction candidates moving forward. An example of this multi-color heatmap is shown in Fig. 3. This heatmap indicates two objects of localization and classification, one green which represents the aluminum and tin can class and one blue which represents polyethylene terephthalate (PETE) plastic class.

Fig. 3. Multi-color Heatmap

7 Object Localization System

First, the 1280×720 image of the conveyor belt of trash is scaled down by 7, to approximately 182 to 103. Every 32×32 sub-image is extracted, resulting in approximately 10,650 images. Each sub-image is predicted on by our one versus all ensemble. Since there are 5 ensembles (one for each class of the one versus all strategy), this returns 5 probability scores of the 5 "one" classes through the process described in Fig. 2. We then check the probabilities of the current image with those directly neighboring it (8 total). If the current image has the highest max probability score out of all its neighbors, we add it to a candidates list. Then, we filter those candidates by removing candidates of the same class that overlap by at least 25% in order to remove redundant predictions. The images with the top 30 max probability scores are kept for the next step. This process is illustrated in Fig. 5.

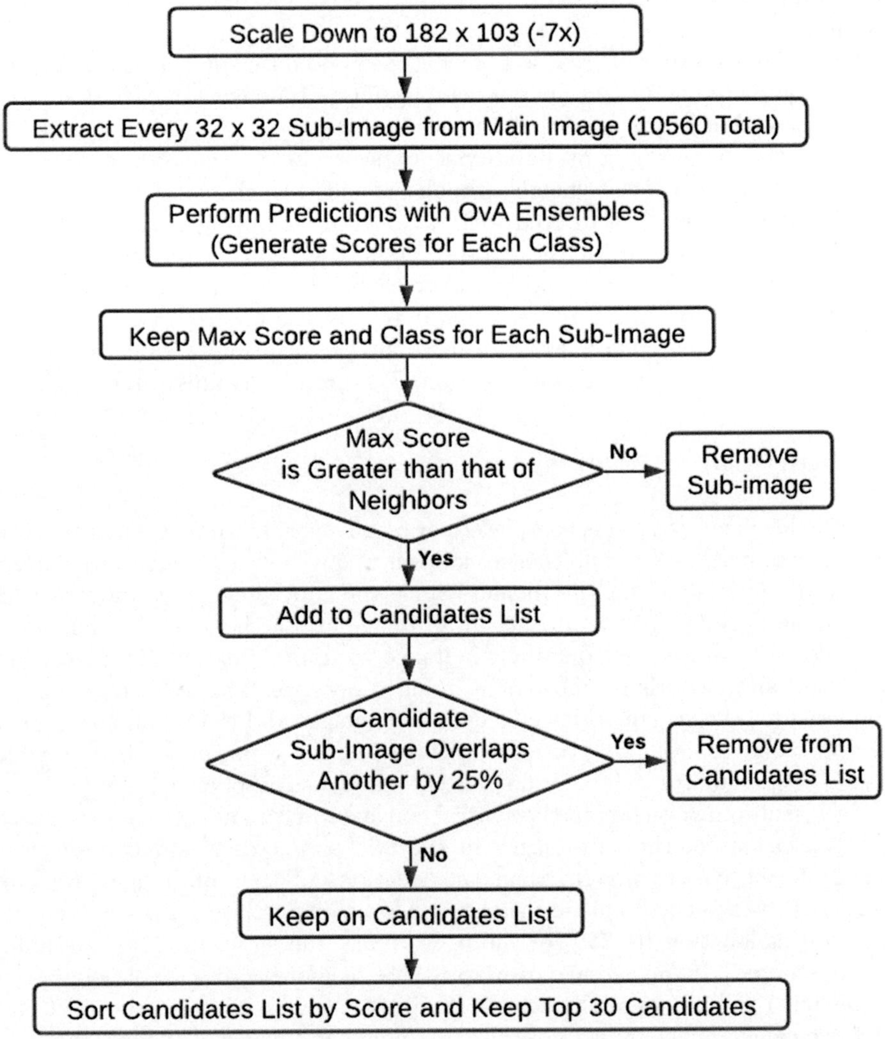

Fig. 4. Low Resolution Localization Flowchart

Next, the location of each sub-image from the previous step is scaled back up (7x) to match the original resolution, and the now 224×224 sub-images are extracted. To account for the imperfect scaling down by 7 of the 1280×820

image, a 7×7 grid of sub-images is extracted with the scaled-up location of the 32×32 sub-images as reference points. This results in 1,470 sub-images of size 224×224. In addition, if an object is predicted in the previous frame, it likely shows up again, so we take those predictions, add 100 pixels to the x location to account for frame movement, and add them into the other 224×224 images to be predicted on again. Not only does this increase confidence in the prediction results, but it removes any false positives that survived the previous frame's filtering.

Next, the following process is used for the candidate sub-images: Each sub-image is predicted on by the OvA ensemble. The sub-image's max probability is compared to that of the other sub-images in the same 7×7 grid. If the current image has the highest max probability score, we add it to a candidates list. Only one image from each 7×7 "neighborhood" is added to the candidates list. The resulting images are then filtered further with the following criteria: The max probability must not be for the "other" class, the probability for the "other" class must be less than 0.4, and the max probability must be greater than or equal to 0.5. The resulting filtered candidates are the finalized results that are stored for the parent image and are used further in a multi-frame analysis, a method described in the following section. This process is illustrated in Fig. 5.

8 Multi-frame Analysis

Once the localization step is completed for a conveyor belt image, the two previous conveyor belt images, and the two following conveyor belt images, we perform a multi-frame analysis on the images. Since the conveyor belt is moving while these images are taken, the objects in an image are farther down or farther up the x-axis in a subsequent or previous image by approximately 100 pixels (+/- 50 pixels), and there is no movement along the y-axis. The following steps are completed for the two previous conveyor belt images and the two following conveyor belt images, adjusting the pixel offset along the x-axis depending on which image is being analyzed (i.e. − 200, − 100, + 100, + 200 pixels).

Every sub-image's class in the central conveyor belt image is cross-referenced with the classes of the sub-images in the two predecessor and two successor images. If the classes match, then the locations of the sub-images are cross referenced. If the central image's sub-image location overlaps a side-image's sub-image offset location by 75% or more, we mark the sub-images as containing the same object. If the average overlap is 90% or more across the 4 neighboring images for a sub-image in the central image, then we keep it for the next step. Next, we calculate the overlap of the remaining sub-images of the central image to filter out images. If two sub-images overlap by 75% or more and have differing classes, we remove the sub-image with the lower probability score. If two sub-images overlap by 50% or more and have the same class, we again remove the sub-image with the lower probability score. Lastly we combine sub-images that overlap (less than 50%) and have the same class to capture an entire object in one larger sub-image if it spans multiple 224×224 sub-images.

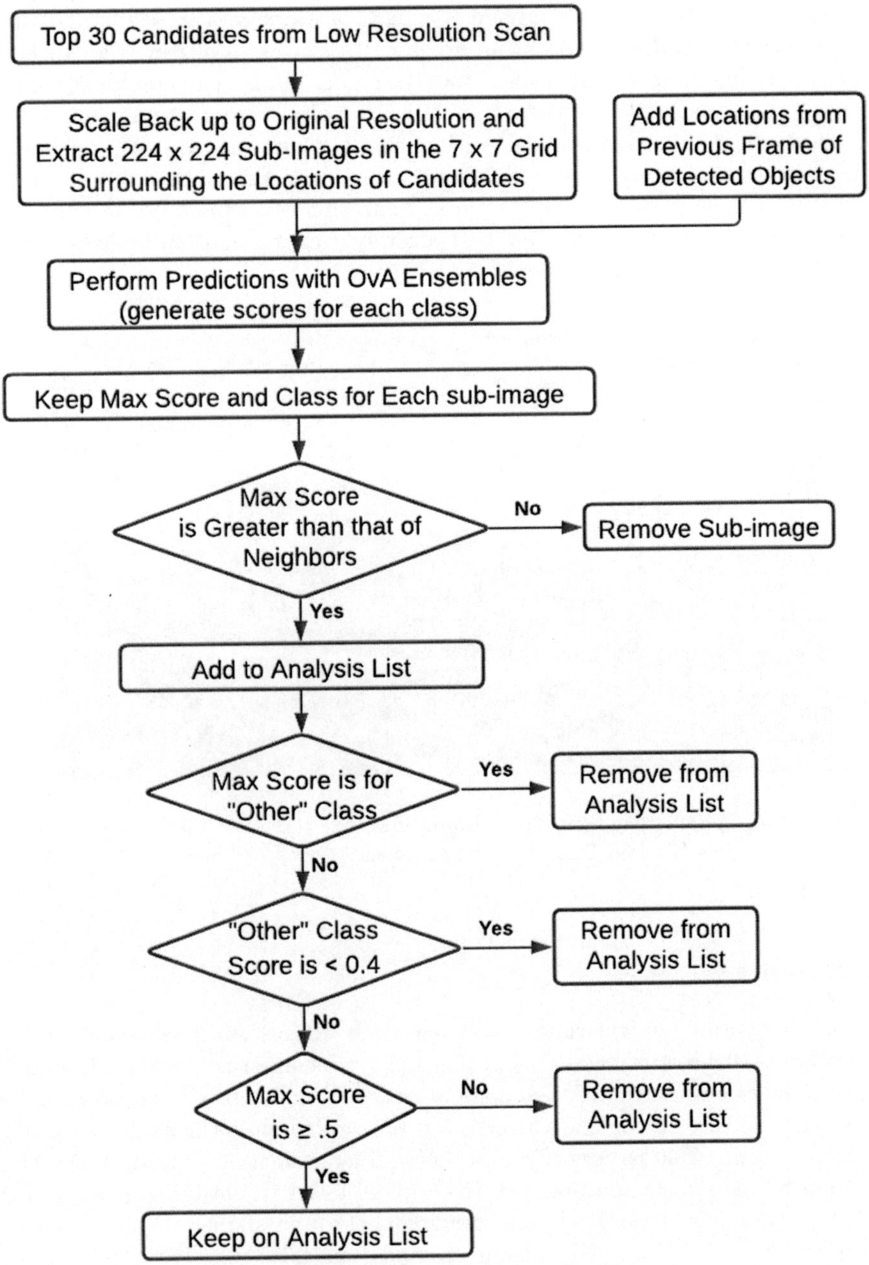

Fig. 5. High Resolution Localization Flowchart

By performing the localization and classification of objects across several conveyor belt images where the x-axis offset is known, we are able to ensemble the results of each individual image to create a more robust result and draw more accurate bounding boxes around these objects. If an object is localized in numerous frames and classified the same, the algorithm is more confident that an object of interest is in fact identified. In turn, if an object is localized in a minority of frames, the ensemble is capable of ignoring the extraneous result through a group consensus. Though the multi-frame analysis increases the computational complexity of the method over the implementation of a single-frame approach, we found that it was a crucial step in accurately drawing bounding boxes around objects of interest and removing false positives.

Fig. 6. Conveyor Belt Image with True Bounding Boxes (Dashed Line), and Our (Solid Line) and Google's (Dotted Line) Detection Boxes

9 Results

On a testing set of 100 consecutive conveyor belt images which contained a total of 89 objects within our 4 classes, our algorithm was able to detect 58 of them for an accuracy of 65.2%. In order to compare this result to the current state of the art, we utilized the Google's AutoML Vision service with the same data. This service was only able to identify 27 objects for an accuracy of 30.3% which is less than half what our method did. In terms of falsely identifying objects, however, the Google's AutoML Vision service performed slightly better than our algorithm. Google's AutoML Vision created four false positive identifications while our method produced seven. To evaluate these results as a single statistic, we utilize the Jaccard Index [23] which is a ratio of the number of correct predictions to the number of incorrect and correct combined. The following table summarizes the results.

Table 1. Summary of Results of Proposed and State of the Art Methods

Predicting Method	True +	False −	False +	Jaccard Index
Novel Method	58	31	7	0.604
Google AutoML Vision	27	62	4	0.290

A sample conveyor belt image is included in Fig. 6. The true bounding box for the objects of interest are contained in the dashed lines, while our bounding (solid line) and Google's AutoML Vision (dotted line) are also shown. The color of the lines specify the class identified. In Fig. 6, Google's AuotML Vision was able to find and classify an aluminum can correctly while our algorithm did the same for a PETE water bottle. An example of a successful object detection by our methodology and unsuccessful by Google's AutoML Vision is shown in Fig. 7.

Fig. 7. Successful Detection by Our Methodology (Solid Line) and Missed by Google with True Bounding Box (Dashed Line)

10 Conclusion and Future Work

In conclusion, the new methodology proposed in this paper for object detection in highly cluttered images performs significantly better than the state of the art. This novel algorithm, that combines the partial least squares algorithm for dimensionality reduction, a newly developed object localization technique, and an ensemble of Extreme Learning Machines for classification, can identify objects in images that even humans struggle to see. We were able to test this novel approach on a data set of conveyor belt images of varying types of trash and achieve an accuracy of 65.2%, more than double Google's AutoML Vision object detection service's accuracy of 30.3%. Though Google's AutoML Vision had fewer false positive detections, the Jaccard Index of both methods highlights that our novel methodology is still superior. AutoML Vision scored a Jaccard Index of 0.290, while our method scored a Jaccard Index of 0.604.

Regarding future work, this new methodology should be tested on more data sets to see if the performance holds across domains. Our data set of conveyor belt images of trash is one in a vast field of imaging brought about by the cheaper cost and higher quality of camera technology. Also, we had to tune numerous parameters within the feature extraction, ELM ensembling, and localization techniques. This took a significant amount of time in our experiment, so the development of automatic parameter tuning would be helpful. By creating a system that could automate the parameter tuning, the algorithm would not only help increase the accuracy on our data set but also make applying this methodology to other data sets easier.

References

1. Singh, A., Pietrasik, M., Natha, G., Ghouaiel, N., Brizel, K., Ray, N.: Animal detection in man-made environments. In: IEEE Winter Conference on Applications of Computer Vision (WACV), vol. 2020, pp. 1427–1438 (2020). https://doi.org/10.1109/WACV45572.2020.9093504
2. Cen, F., Zhao, X., Li, W., Wang, G.: Deep feature augmentation for occluded image classification. Pattern Recognit. **111**, 107737 (2021). https://doi.org/10.1016/j.patcog.2020.107737
3. Liu, N., Wang, H.: Evolutionary extreme learning machine and its application to image analysis. J. Signal Process. Syst. **73**, 73–81 (2013). https://doi.org/10.1007/s11265-013-0730-x
4. Li, J., Zhao, X., Li, Y., Du, Q., Xi, B., Hu, J.: Classification of hyperspectral imagery using a new fully convolutional neural network. IEEE Geosci. Remote Sens. Lett. **15**(2), 292–296 (2018). https://doi.org/10.1109/LGRS.2017.2786272
5. Nagarajan, B.: Object classification in static images with cluttered back ground using statistical feature based neural classifier. Asian J. Inform. Technol. **7**, 162–167 (2008)
6. Huang, G., Zhu, Q.-Y., Siew, C.-K.: Extreme learning machine: a new learning scheme of feedforward neural networks. In: 2004 IEEE International Joint Conference on Neural Networks (IEEE Cat. No.04CH37541), Vol. 2, pp. 985–990 (2004)
7. Huang, G., Zhu, Q., Siew, C.: Extreme learning machine: theory and applications. Neurocomputing **70**(1), 489–501 (2006)
8. Huang, G., Zhou, H., Ding, X., Zhang, R.: Extreme learning machine for regression and multiclass classification. IEEE Trans. Syst. Man Cybern. Part B (Cybern.) **42**(2), 513–529 (2012)
9. Huang, G., Chen, Y.-Q., Babri, H.A.: Classification ability of single hidden layer feedforward neural networks. IEEE Trans. Neural Netw. **11**(3), 799–801 (2000)
10. Li, Z., Ratner, K., Ratner, E., Khan, K., Bjork, K.-M., Lendasse, A.: A novel ELM ensemble for time series prediction. In: Proceedings of ELM 2018, vol. 11, pp. 283–291. Springer, Cham (2019)

11. Khan, K., Ratner, E., Ludwig, R., Lendasse, A.: Feature bagging and extreme learning machines: machine learning with severe memory constraints. (2020) pp. 1–7. https://doi.org/10.1109/IJCNN48605.2020.9207673
12. Guillen, A., Herrera, L., Rubio, G., Pomares, H., Lendasse, A., Rojas, I.: New method for instance or prototype selection using mutual information in time series prediction. Neurocomputing **73**, 2030–2038 (2010). https://doi.org/10.1016/j.neucom.2009.11.031.
13. Montesino Pouzols, F., Lendasse, A., Barros, A.: Autoregressive time series prediction by means of fuzzy inference systems using nonparametric residual variance estimation. Fuzzy Sets Syst. 471–497 (2010). https://doi.org/10.1016/j.fss.2009.10.018
14. Roshan, S., Miche, Y., Akusok, A., Lendasse, A.: Adaptive and online network intrusion detection system using clustering and extreme learning machines. J. Franklin Inst. **355**. https://doi.org/10.1016/j.jfranklin.2017.06.006
15. Haykin, S.: Neural Networks: A Comprehensive Foundation. Prentice Hall PTR Upper Saddle River, NJ, USA (2004)
16. Huang, G.: What are extreme learning machines? Filling the gap between Frank Rosenblatt's dream and John von Neumann's puzzle. Cogn. Comput. **7**(3), 263–278 (2015)
17. Yu, H., Yuan, Y., Yang, X., Dan, Y.: A dynamic generation approach for ensemble of extreme learning machines. In: Zeng, Z., Li, Y., King, I. (eds.) Advances in Neural Networks - ISNN 2014, pp. 294–302. Springer International Publishing, Cham (2014)
18. Bisong, E.: An Overview of Google Cloud Platform Services, pp. 7–10. Apress, Berkeley, CA (2019)
19. Simonyan, K., Zisserman, A.: Very deep convolutional networks for large scale image recognition. arXiv 1409.1556
20. Liu, Z., Wu, J., Fu, L., Majeed, Y., Feng, Y., Li, R., Cui, Y.: Improved kiwifruit detection using pre-trained vgg16 with rgb and nir information fusion. IEEE Access **8**, 2327–2336 (2020). https://doi.org/10.1109/ACCESS.2019.2962513
21. Rosipal, R., Kramer, N.: Overview and recent advances in partial least squares. In: Saunders, C., Grobelnik, M., Gunn, S., Shawe-Taylor, J. (eds.) Subspace Latent Structure and Feature Selection, pp. 34–51. Springer, Berlin Heidelberg, Berlin, Heidelberg (2006)
22. Schwartz, W.R., Kembhavi, A., Harwood, D., Davis, L.S.: Human detection using partial least squares analysis. In: 2009 IEEE 12th International Conference on Computer Vision, pp. 24–31 (2009). https://doi.org/10.1109/ICCV.2009.5459205
23. Jaccard, P.: The distribution of the flora of the alpine zone. New Phytol. **11**, 37–50 (1912)

Extreme Learning Machines for Offline Forged Signature Identification

Zhen Li[1(✉)], Leonardo Espinosa-Leal[1], Amaury Lendasse[2], and Kaj-Mikael Björk[1]

[1] Department of Business Management and Analytics, Arcada University of Applied Sciences, Jan-Magnus Janssons plats 1, 00560 Helsinki, Finland
{lizhen,leonardo.espinosaleal,kaj-mikael.bjork}@arcada.fi
[2] University of Houston, 77004 Houston, TX, USA
alendass@Central.UH.EDU

Abstract. Signature verification has a crucial role in individual authentication process, and forged signatures can cause vast damages in all the fields. An important aspect for signature verification is that there are millions of credit card transactions providing signatures; therefore, speedy checks are required. We propose using extreme learning machines (ELM) to help with the task. We demonstrate the effectiveness and efficiency of using ELM detecting forged signatures. The proposed method reports 0.27% equal error rate, compared to recent 0.88% reported results. Linear neuron type outperforms other neuron types in this task, as it converges the fastest regardless of feature extractors used. However, the best performance was still reached by using RELU with SigNet as feature extractor.

Keywords: Signature forgery · Extreme learning machines · Signal processing · Writer dependent

1 Introduction

Biometric authentication methods have always been a major player in personal authentication. Improvements in hardware and software in mobile devices have converted Face ID and fingerprint as a dominant means for authentication. However, despite other advances in user identification, in other applications such as legal documents and credit card authentication, signature or electronic signature is the to-go method [11].

In signature identification, there are two different setup types: offline and online. An offline signature will only have the signature's visual information; an online signature will have extra information such as how long it took to sign and how the pressures are for each signature parts/pixels.

Due to the importance of signature identification, many fraudulent attempts are trying to forge signatures. If a forgery is produced by a skillful forger trained

K.-M. Björk (Ed.): ELM 2021, PALO 16, pp. 24–31, 2023.
https://doi.org/10.1007/978-3-031-21678-7_3

on target signature's static copies, it is challenging to distinguish in an offline setup. This is mainly because the handwriting style of a person is dynamic and non-stationary. A person's signature evolves, especially when the time gap is significant. With a limited amount of information in an offline setup, the forged signatures look like real ones.

In this paper, we attempt to tackle the problem as mentioned above, i.e., distinguish forgeries from genuine signatures offline, thanks to extreme learning machine (ELM).

2 Related Works

Many works have been done in signature identification. Some studies have been using autoencoders to verify signatures [1,18]. Autoencoder has encoder and decoder components. The encoder part tries to project the given data to a much smaller latent space. Then, the decoder tries to recover the original input data from the projected latent representation. Therefore, the encoder and the decoder will capture the patterns in the data, and forgeries will have a worse recovered quality compared to authentic signatures.

On the other hand, researchers are trying to use Convolution Neural Networks (CNN) to extract features from handwritten signatures [8–10,17] Using CNN to extract features has been proven to be effective [9,10]. The two neural architectures proposed in the papers are named SigNet [10] and SigNet-SPP [9] (see Fig. 1), with SPP referring to spatial pyramid pooling. Previous works have shown that ELM can efficiently and accurately verify static signatures using the features extracted by those neural network feature extractors [3,6].

ELM is a fast universal approximator. Its straightforward single-layer structure guarantees its computation efficiency. Besides, it can fit non-linear functions as long as initial weights are generated randomly and adequately. In this paper, we use the ELM implementation called Scikit-ELM [2]. The package has been developed to be compatible with the popular scikit-learn package [16].

ELM has been shown to work well for vision tasks. Chen et al. showed that given limited amount of training samples, ELM outperforms other known approaches in hyperspectral image classification [5]. Mohammed et al. has shown the effectiveness of ELM along with bidirectional two-dimensional principal component analysis on human face recognition task [15]. Ma et al. showed ELM can detect falling action from bending, sitting, squatting, walking, and lying actions in video clips with high sensitivity and fair accuracy [14].

This work builds upon a previous model where the authors trained an ELM for independent writer signatures [6], which means that they did not focus on detecting forgeries of a specific user. In this paper, we will focus on the writer dependent forgery classification task using the standard Equal Error Rate (EER) as the metric.

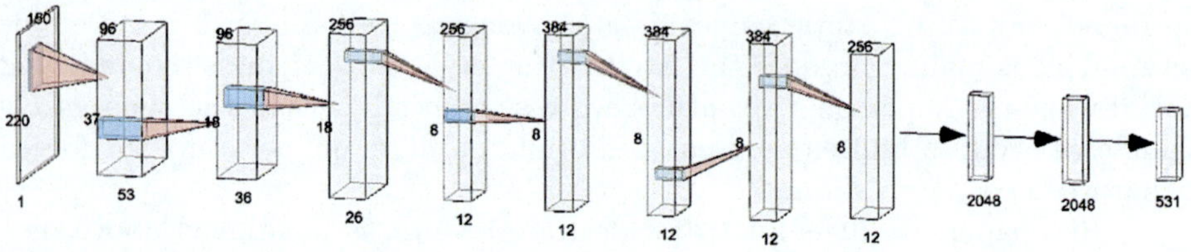

Fig. 1. Neural architecture of Signet[10] Through multiple convolutional layers, the feature information is extracted to a vector.

3 Research Methodology

In this work, we use the MCYT-75 dataset [7]. There are 75 writers in the dataset, with 15 forged and 15 genuine signatures. This dataset is tiny compared to the available data for research, but it works as a benchmark for evaluating performance when data is limited, as real use cases. We use Inception21k [12, 13], SigNet, SigNet300 and SigNet600 to extract features separately from the signature images.

All the features are in the vector format and will be used as the training data for an ELM classifier. We also concatenate all the features extracted by the four extractors mentioned before as a training start too. Each writer will have its classifier trained on its forged and genuine signature samples. So, we will have 75 classifiers in total. Given this writer-dependent strategy, we can focus on forgeries for each writer.

We randomly reserved 30% of the forged signatures and genuine signatures for each writer as a test dataset, as shown in Fig. 2. When training, we applied five-fold cross-validation for all the 75 classifiers. Finally, the model is trained with all the training data, and we report the model's EER on test data reserved previously.

4 Results and Discussion

From the experiments on the MCYT-75 dataset, the result looks promising. The results are compiled in Tables 1 and 2. The changes of EER against the number of neurons for different neuron types are presented in Fig. 3 As we can see from both the tables and figures, the EER decreases fast with the increase in neuron number regardless of the neuron types or feature extractors. Furthermore, the performance reached a maximum of around 128 neuron numbers, which is similar to the results reported before [6]. This is mainly due to the size limit of the MCYT-75 dataset. SigNet related feature extractors share similar performance when the number of neurons and neuron types are given. In the meanwhile, linear neuron type outperforms other neuron types in this task.

Regardless of the feature extractor, the EER using linear neurons is significantly better than other neuron types. This gap is apparent when the neuron number is around 32. The EER of linear neurons is roughly half of that using

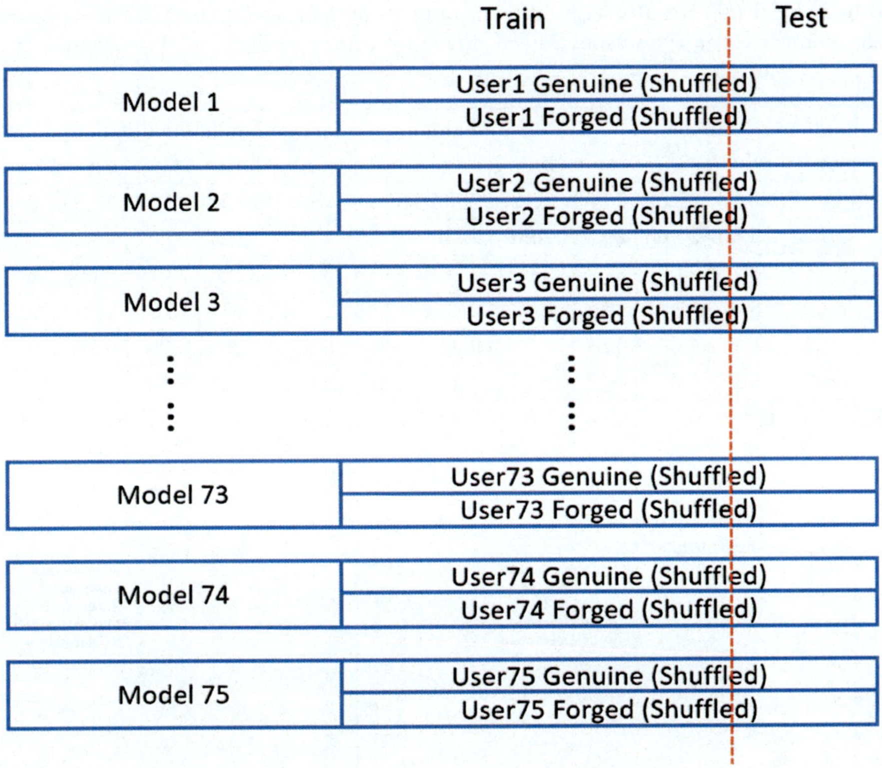

Fig. 2. Training strategy and data split

Table 1. EER (%) Results with different number of neurons for the MCYT-75 dataset with various neuron type using Inception21k and all combined - including Inception21k, SigNet, SigNet300dpi, and SigNet600dpi - as feature extractors.

	Inception21k				Combined			
n	lin	relu	sig	tanh	lin	relu	sig	tanh
4	25.07	26.93	31.20	34.67	23.47	33.07	34.67	32.27
8	22.67	24.00	31.47	32.80	18.67	24.53	27.47	31.73
16	25.33	26.13	26.93	28.53	25.07	22.93	31.73	28.00
32	14.93	29.33	29.87	32.27	9.33	19.20	21.87	26.40
64	7.47	12.53	12.27	17.33	1.60	4.53	8.00	11.737
128	3.47	5.87	3.73	4.53	0.27	0.80	2.40	3.20
256	4.27	4.53	4.80	5.60	0.80	0.80	0.27	0.53

Table 2. EER (%) Results with different number of neurons for the MCYT-75 dataset with various neuron type using SigNet, SigNet300dpi, and SigNet600dpi separately as feature extractors.

	SigNet				SigNet300dpi				SigNet600dpi			
n	lin	relu	sig	tanh	lin	relu	sig	tanh	lin	relu	sig	tanh
4	18.67	23.73	28.80	28.53	20.27	30.67	30.13	28.27	18.13	18.93	29.33	30.67
8	16.53	22.40	25.60	29.33	26.67	29.33	26.67	26.93	16.27	26.67	28.80	28.27
16	17.60	23.73	24.53	27.73	28.00	27.47	28.53	32.27	22.67	23.73	26.93	30.13
32	8.00	18.67	17.87	26.13	9.33	21.33	17.60	21.87	5.87	15.73	14.93	20.00
64	1.87	7.47	4.00	6.67	3.73	9.33	4.80	8.53	3.20	8.00	3.73	4.27
128	0.53	1.07	1.07	1.87	2.13	3.47	2.40	2.67	1.07	2.13	1.87	2.40
256	0.53	0.27	0.53	0.80	1.07	1.60	1.33	1.07	0.80	1.33	0.80	1.07

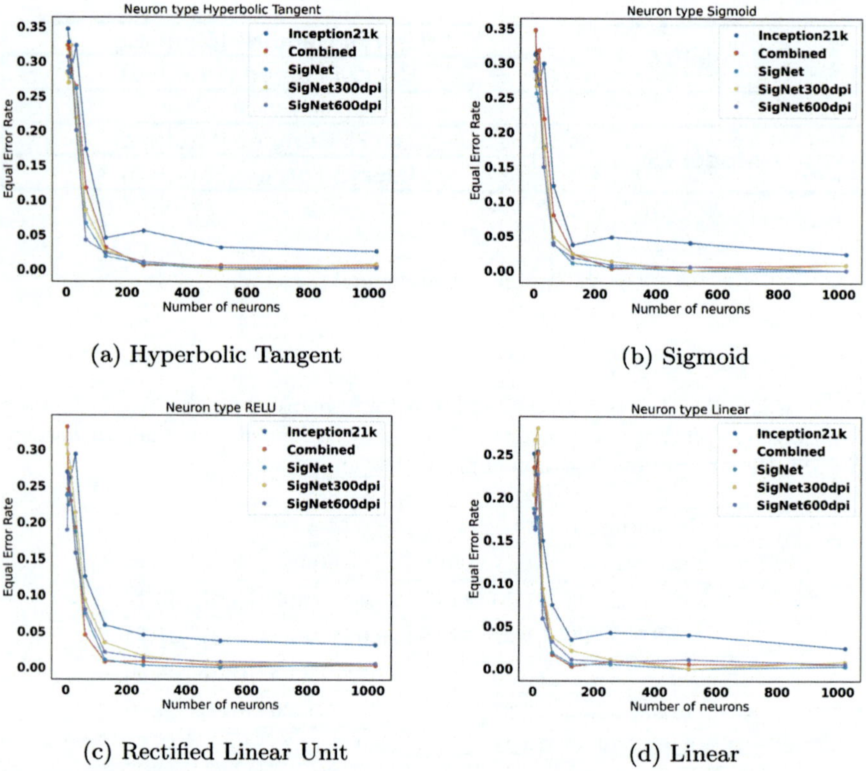

(a) Hyperbolic Tangent

(b) Sigmoid

(c) Rectified Linear Unit

(d) Linear

Fig. 3. EER against number of neurons for different feature extractors. Combined refers to the features including all the features extracted by the four extractors.

other neuron types compared within specific feature extractors. As discussed in previous works [6] SigNet based feature extractors are better fitted for this signature verification task as they were explicitly trained, whereas Inception21k was designed and trained for a more general vision task. However, when using linear neurons, the EER obtained using Inception21k as a feature extractor is roughly similar to that obtained using SigNet family feature extractors with non-linear neurons. This linear neuron outperforming other neuron types might be because of the writer-dependent setup. All the features can be and should be compared directly without too much non-linearity introduced.

From Tables 1 and 2, we can see that the combined features obtained by concatenating the features extracted by different feature extractors are not helping with the EER much, especially when the number of neurons is small. This might be because of the data size limitation. In addition, the writer-dependent setup might require less on the varieties of features but focus more on the feature variance. Therefore, the change within the same feature is more important than capturing different features when neuron numbers are small. In addition, a point worth noticing is that ELM copes well with the collinearity issues because the combined features lead to similar EER performance as individual ones.

Despite the combined features not having much obvious advantage on EER especially when neuron numbers are small, the best EER performance, 0.27% was achieved by 128 linear neurons with combined features. The same performance was reached with 256 linear neurons with SigNet feature extractor. Combined features need less neurons as well as linear neurons to reach the best performance might indicate that the combined features capture some extra information that's vital for the last performance boost. However, this extra information can still be found by ELM with more neurons with single feature extractor. The 0.88% EER was reported in a recent study[4]. As can be seen from both tables, the proposed method also reached 0.80% EER, the improvement from 0.80% to 0.27% might be because of the extra information from combined features or fitting capability from ELM.

5 Conclusion and Future

This paper presented a set of experiments using ELM combined with different CNN-based feature extractors to identify forged signatures from genuine signatures in a writer-dependent setup. In general, the performance in terms of EER is comparable to other recent signature verification works [4]. We found that linear neuron type is much better in the task than other popular neuron types regardless of the feature extractors. In addition, in writer dependent setup, more features will not help much with EER performance. In terms of future work, we would like to apply the method to a more extensive dataset to see if training data size limits the performance.

Acknowledgments. The authors wish to acknowledge CSC – IT Center for Science, Finland, for computational resources.

References

1. Ahrabian, K., Babaali, B.: On usage of autoencoders and siamese networks for online handwritten signature verification. corr. arXiv preprint arXiv:1712.02781 (2017)
2. Akusok, A., Espinosa Leal, L., Björk, K.M., Lendasse, A.: Scikit-elm: an extreme learning machine toolbox for dynamic and scalable learning. In: International Conference on Extreme Learning Machine, pp. 69–78. Springer (2019)
3. Akusok, A., Espinosa Leal, L., Björk, K.M., Lendasse, A., Hu, R.: Handwriting features based detection of fake signatures. In: The 14th PErvasive Technologies Related to Assistive Environments Conference. pp. 86–89 (2021)
4. Berkay Yilmaz, M., Ozturk, K.: Hybrid user-independent and user-dependent offline signature verification with a two-channel cnn. In: Proceedings of the IEEE Conference on Computer Vision and Pattern Recognition Workshops. pp. 526–534 (2018)
5. Chen, C., Li, W., Su, H., Liu, K.: Spectral-spatial classification of hyperspectral image based on kernel extreme learning machine. Remote Sens. **6**(6), 5795–5814 (2014)
6. Espinosa-Leal, L., Akusok, A., Lendasse, A., Björk, K.M.: Extreme learning machines for signature verification. In: International Conference on Extreme Learning Machine, pp. 31–40. Springer (2019)
7. Fierrez-Aguilar, J., Alonso-Hermira, N., Moreno-Marquez, G., Ortega-Garcia, J.: An off-line signature verification system based on fusion of local and global information. In: International workshop on biometric authentication, pp. 295–306. Springer (2004)
8. Gideon, S.J., Kandulna, A., Kujur, A.A., Diana, A., Raimond, K.: Handwritten signature forgery detection using convolutional neural networks. Procedia Comput. Sci. **143**, 978–987 (2018)
9. Hafemann, L.G., Oliveira, L.S., Sabourin, R.: Fixed-sized representation learning from offline handwritten signatures of different sizes. Int. J. Doc. Anal. Recogn. (IJDAR) **21**(3), 219–232 (2018)
10. Hafemann, L.G., Sabourin, R., Oliveira, L.S.: Learning features for offline handwritten signature verification using deep convolutional neural networks. Pattern Recogn. **70**, 163–176 (2017)
11. Henniger, O., Muramatsu, D., Matsumoto, T., Yoshimura, I., Yoshimura, M.: Signature Recognition, Overview, pp. 1–4. Springer, US, Boston, MA (2009)
12. Leal, L., Akusok, A., Björk, K.: Classification of websites via full body renders. In: Proceedings of the ELM (2019)
13. Leal, L.E., Björk, K.M., Lendasse, A., Akusok, A.: A web page classifier library based on random image content analysis using deep learning. In: Proceedings of the 11th PErvasive Technologies Related to Assistive Environments Conference, pp. 13–16 (2018)
14. Ma, X., Wang, H., Xue, B., Zhou, M., Ji, B., Li, Y.: Depth-based human fall detection via shape features and improved extreme learning machine. IEEE J. Biomed. Health Inf. **18**(6), 1915–1922 (2014)
15. Mohammed, A.A., Minhas, R., Wu, Q.J., Sid-Ahmed, M.A.: Human face recognition based on multidimensional pca and extreme learning machine. Pattern Recogn. **44**(10–11), 2588–2597 (2011)
16. Pedregosa, F., Varoquaux, G., Gramfort, A., Michel, V., Thirion, B., Grisel, O., Blondel, M., Prettenhofer, P., Weiss, R., Dubourg, V., et al.: Scikit-learn: machine learning in python. J. Mach. Learn. Res. **12**, 2825–2830 (2011)

17. Poddar, J., Parikh, V., Bharti, S.K.: Offline signature recognition and forgery detection using deep learning. Procedia Comput. Sci. **170**, 610–617 (2020)
18. Prajapati, P.R., Poudel, S., Baduwal, M., Burlakoti, S., Panday, S.P.: Signature verification using convolutional neural network and autoencoder. J. Inst. Eng. **16**(1), 33–40 (2021)

Randomized Model Structure Selection Approach for Extreme Learning Machine Applied to Acid Sulfate Soil Detection

Anton Akusok[1(✉)], Kaj-Mikael Björk[1], Virginia Estévez[1], and Anton Boman[2]

[1] Arcada University of Applied Sciences, Helsinki, Finland
anton.akusok@arcada.fi

[2] Geological Survey of Finland, FI-67101 Kokkola, Finland

Abstract. This paper investigates the performance of Extreme Learning Machine in acid sulfate soil classification task. A special focus is made on a robust model structure selection by randomized parameter search. The resulting model structures are compared using two datasets about the same phenomenon but of a very different nature: one with hand-crafted scalar features, and the other with 9×9 pixels image patches instead of features. The results highlight the flexibility of the presented model selection approach.

Keywords: Extreme learning machine · Classification · Model structure selection · Acid Sulfate Soil

1 Introduction

Acid sulfate (AS) soils are common types of soils in the coastal areas of northern European countries, Finland being the one with larger concentrations. Drainage of the upper layer of this type of soil due to agricultural and infrastructure uses can cause ecological damages [6]. This is because drainage can lead to oxidation of iron sulfides, which lead to formation of sulfuric acid, which may wash away, potentially causing significant environmental damage [5]. These damages can be minimized by limiting drainage depth [3]. Thus, predicting the exact location of AS soils becomes an important research problem with valuable practical applications.

So far, the localization of AS soils has been made with traditional methods, which consist of laboratory testing of soil samples collected in the field by geologists. Soil samples are classified as AS soil or non-AS soils [7]. This is a very expensive and time-consuming process. Machine learning methods can use the existing soil samples for training the models and predicting AS soils. These techniques can speed up the process and lower mapping costs. [9] Machine learning techniques such as artificial neural networks [4], fuzzy logic [5], or random forest and gradient boosting [8,9] have been successfully used for predicting AS soils.

K.-M. Björk (Ed.): ELM 2021, PALO 16, pp. 32–40, 2023.
https://doi.org/10.1007/978-3-031-21678-7_4

This study considers an Extreme Learning Machine (ELM) method for detecting AS soils, with a specific focus on the model structure selection. Machine learning models include hyperparameters - special variables whose optimal values cannot be computed by the model training procedure, and has to be specified manually before the training begins. Usually hyperparameters relate to the model behavior, like different types of neural network layers, numbers of neurons or connections between them. Hyperparameters allow the same model to adapt to different configurations of training data. This is important for AS soil detection data that is very limited in size and comes in two different formats.

Good values of hyperparameters can be found by a grid search, usually with the cross-validation error as an optimization metric. The large number of hyperparameters makes the model very flexible with regard to different kinds of input data, but finding good values for those prove challenging as the grid search complexity is exponential in the number of hyperparameters. Randomized search addresses the exponential complexity issue by evaluating model performance on random combination of hyperparameter values, suggesting good values with a large number of trials. Efficient utilization of randomized hyperparameter search requires a fast model, and a way to extract good point values from the collected trial data. The fast model requirement is addressed by using ELM, and this paper proposes a convenient way of merging trial data for hyperparameter optimization.

The next sections introduce the datasets with two different views on the same problem, and an Extreme Learning Machine method with the randomized parameter search methodology based on modern programming libraries. The results section compares the performance of the proposed method with the alternatives, and examines the effect of wide-range parameter search on the final model structure. The conclusions section summarizes the findings.

2 Datasets

The area of interest in this paper is called Virolahti, which is located in south-eastern Finland. The datasets used are the same as those created in a previous work [8]. There are two different types of data, the point observations and the environmental layers. The point observations (PO) are soil samples taken in the field by researchers from the Geological Survey of Finland (GTK). These PO are classified as AS soil or not.

The environmental covariates are raster data, usually obtained from remote sensing data. Different types of covariates are considered: digital elevation model (DEM), terrain, geophysics, and quaternary. These environmental layers have been created from data of different sources, the quaternary and geophysics layers from GTK, the DEM from National Land Survey of Finland (NLS), and the terrain layer is derived from the DEM. In this work, five environmental layers are considered: DEM, slope, Quaternary geology map, which shows 12 classes of materials 1, and the real and imaginary electromagnetic components. All these layers show valuable information for the characterization of the soils.

Table 1. Types of soils

Class	Soil type
1	Bedrock outcrop
2	Bedrock at surface
3	Till
4	Fine sand to gravel
5	Cobbles and boulders
6	File-silt
7	Clay
8	Gyttja
9	Unmapped area
10	Water
11	Fine silt with low humus
12	Thick peat deposits

The information from the environmental layers can be used as scalar values or images. Thus, two datasets are used, the first one scalar and the second one of images. In the scalar dataset, the values of the covariates for the point observations are considered. The second image patch dataset consists of 9×9 pixel images of the covariates, centered at the location of each point observation [8]. The size of the pixel is 50 m x 50 m.

The dataset includes a total of 186 observations, with a fixed split into 148 training data samples and 38 test data samples. The quaternary geological map is presented in one-hot encoding, and it has 9 classes of materials present in the dataset out of 12. This results in $4 + 9 = 13$ features for scalar dataset, and $(4*9*9)+9 = 333$ features for the visual maps of environmental covariate layers dataset.

3 Methodology

3.1 Extreme Learning Machine

In this study, an Extreme Learning Machine method is considered for detecting AS soils. It is a randomized neural network with a single hidden layer that scales from very small to the largest tasks, has very fast training, and supports a broad set of configuration options including sparsely connected or pairwise function neurons. Model structure selection from a large variety of configurations is performed by a randomized parameter search combined with cross-validation in a convenient Scikit-Learn[1] implementation. Randomized search is the only

[1] https://scikit-learn.org/stable/modules/grid_search.html#randomized-parameter-optimization.

feasible option with a multitude of different options, as grid search space on model parameters quickly become too large as the number of possible parameters grows.

Extreme Learning Machine (ELM) [11] features a randomly generated and fixed weights connecting input and hidden layer neurons. Because only the output weights are unknown, their solution is obtained efficiently in a closed form [1]. The non-linear transformation function in a hidden layer gives non-linear properties to the whole method, allowing it to perform with arbitrary data on par with other general machine learning algorithms [10].

An ELM algorithm places no restrictions on a form or type of hidden layer neurons, except that their output values should be finite. This allows for an extremely wide range of possible hidden neurons - including sparse connections, pairwise function neurons (also known as Radial Basis Functions), different non-linear transformation functions; and any of their combinations. It is possible to add extra special hidden neurons that simply copy input features, adding a linear part into an otherwise non-linear model. Other ELM hyperparameters include the number of hidden neurons, and the regularization constraints for the output weights computation. An overview of ELM structure, and possible parameter options are shown on Fig. 1.

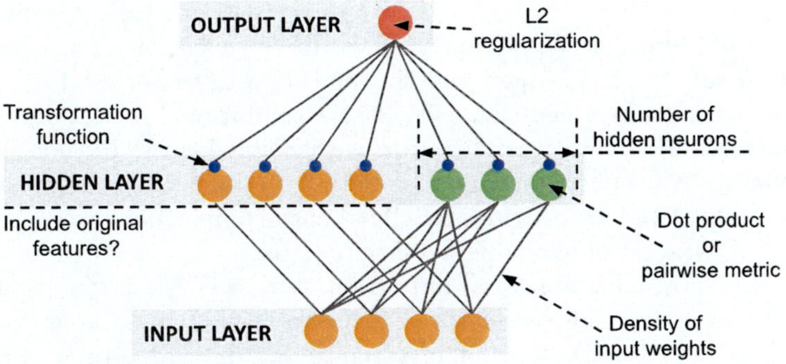

Fig. 1. Extreme Learning Machine and its hyperparameters.

3.2 Randomized Model Structure Selection

The variability of ELM as a machine learning method should allow it to adapt to any kind of dataset, given a sufficiently good model structure selection approach. The approach should produce an optimal combination of hyperparameters.

Typically, model hyperparameters are tested on a parameter grid, taking a range of values for each one and validating all their combinations - for instance with cross-validation. A drawback of such approach is its exponential scaling of runtime with the number of parameters. Testing n parameters with m values each on a k-fold cross-validation requires km^n model evaluations. This quickly becomes infeasible as the number of hyperparameters goes beyond 2 or 3.

Randomized parameter search alleviates the problem of infeasibly large parameter spaces. It cannot possibly test all combinations of parameters to find the precise best values, but it quickly discriminates between well-performing and poorly-performing ranges of model parameters. ELM model is not very sensitive to the precise values of hyperparameters due to its random nature, making it sufficient to obtain an approximate range of well-performing values.

A unique feature of randomized parameter search is the discovery of well-performing combinations of multiple parameters outside of their typical examined values. Combined with a highly configurable model like ELM, the approach is a good fit for phenomenon analysis from different data modalities, as shown in the next section.

4 Results

This work will examine how well AS soil samples can be classified with the ELM model and the different datasets.

Studies are performed on a dataset taken with a pre-defined split between training/validation and test sets. This enables comparison of the proposed ELM method to a recent thesis work [8] that examines the same task in depth. In this previous work, the scalar dataset has been used for conventional machine learning methods: random forest, gradient boosting and support vector machines, whereas the dataset with images has been used for a convolutional neural network.

Randomized parameter search is run with Scikit-Learn[2] Python ecosystem that provides a convenient interface for both experimentation and the analysis of results. Scikit-ELM [2] provides a Scikit-Learn compatible implementation of ELM, with a wide set of available hyperparameters.

Data value normalization is an important step in ELM, as well as any other model that combines multiple features in its evaluation (a notable exception is Decision Tree-based methods that investigate one data feature at a time, and are unaffected by difference in value ranges between the features). The model pipeline includes a robust data scaler before an ELM, that normalizes data value range using an interquartile range between the 1st and the 3rd quartiles. In Scikit-Learn framework, the scaler correctly processes data using the training folds of k-fold cross-validation, and applies the computed transformation to the validation fold.

Randomized Search Setup

Model performance is evaluated on a 3-fold cross validation, with experiments running a couple hours on a desktop PC saving their results. The final hyperparameter values are defined as median values, or most common values for multiple

2 https://scikit-learn.org.

choice features, among the 5% trials with the lowest cross validation error. An identical approach is used for both datasets.

Hyperparameters and their range of values are listed below:

Number of neurons: Number of hidden layer neurons in ELM model. More neurons increase learning capacity and a risk of overfitting. Sampled uniformly from an interval $[3, 400]$.

L2-regularization term α**:** Regularization term for the output layer, reduces model overfitting. Sampled uniformly *on a logarithmic scale* from an interval $[10^{-4}, 10^{4}]$.

Transformation function: A transformation function applied to hidden layer output, adds non-linearity to ELM model as a whole. Possible choices are a sigmoid, a hyperbolic tangent, a rectified linear units, or no transformation function at all (an identity function).

Pairwise metric: An optional pairwise metric that replaces dot product in hidden neuron computation with a distance function computed from an input data sample and a random vector of the same length stored in a hidden neuron. Possible distance functions are l2-distance, l1-distance, cosine distance, and no distance function (a standard dot product).

Density: Density of the random connection weights between input and hidden layers. Density below 1.0 removes the corresponding portion of connection weights, similar to dropout rate in neural networks training. The effect allows hidden neurons to focus on different feature combinations in the input data, instead of every hidden neuron being connected to all the input data features at once. Values are sampled uniformly from an interval $[0.01, 1]$.

Include original features: An optional choice of concatenating the original input data features as extra features to the hidden layer output, before computing the output weights. Adding the original features to non-linear hidden layer features includes a linear part in ELM, allowing it to model any linear dependencies in the data directly without a non-linear approximation. The drawback is an increased number of features that may lead to overfitting, especially with high original data dimensionality of noisy data features. Possible options are true or false.

Results for the Two Different Datasets

The results of ELM classifier are shown in Fig. 2, in comparison to other traditional models trained in [8]. ELM includes randomly generated weights, so the model with best hyperparameters is re-trained 1000 times, and average test results are reported.

ELM with point data performs on par with the alternatives in non-AS soil detection. It performs worse in AS soil detection, but the ELM results are improved by the inclusion of visual maps of environmental covariate layers. Performance with visual maps is still limited, probably due to the very limited training data size for the high-dimensional input data. Interestingly, ELM per-

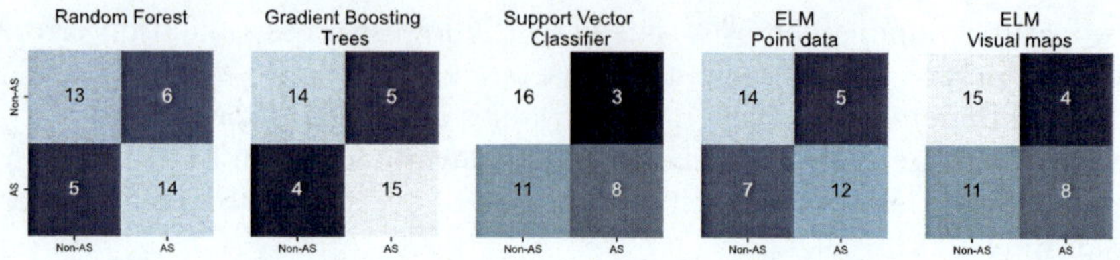

Fig. 2. Confusion matrices for test set predictions of ELM in comparison with three other methods in [8].

formance is identical to SVM, suggesting there might be a general local optimum discovered by both an SVM and an ELM.

4.1 Comparison of Discovered Model Structures

An interesting observation can be done on the discovered model structures from the two very different data sources. The summary of optimal hyperparameter values is given in Table 2.

Table 2. Comparison of optimal values for model hyperparameters discovered by random parameter search.

Parameter	Scalar features dataset	Image patches dataset
Number of neurons	22	15
L2-regularization α	$10^{1.0}$	$10^{1.9}$
Transformation function	ReLU	Sigmoid
Pairwise metric	L1-distance	none
Input weights density	0.74	0.09
Include original features	no	no

ELM model adapted differently to the two data types, despite using exactly the same point observations for soil type classification. On point observation dataset, the model decides to use pairwise distance function in the hidden neurons. The number of hidden neurons is relatively small and the regularization term of the output that prevents overfitting to the training set is substantially larger than its common value of $\alpha = 10^{-5}...10^{-3}$ that is enough to prevent numerical computation problems.

On image patches the model uses an even smaller number of hidden neurons, in addition to very sparse hidden layer weights and a larger L2 regularization term. This points to problems with overfitting to the data, that the model counters by reducing its learning capacity with less weights to train, and a higher model regularization. This is an expected outcome for the training dataset of

148 samples and 333 features - there are twice more features than data samples. The selected hyperparameter values fit well to the purpose of reducing the overfitting.

5 Conclusions

This work investigates the use of Extreme Learning Machines with randomized model structure selection for AS soil detection in southeastern Finland. A universal approach is applied on two datasets, where the information of the environmental layers is given in two very different ways. In the first one, the features are the values of the covariates for the point observations. While the second one consists of images of the environmental covariates centered at the point observations.

An ELM model achieved comparable performance to the available alternative methods like SVM or ensemble of decision tree-based methods, although it suffered from overfitting on a small training set that is visible in the selected values of ELM hyperparameters.

Randomized model structure selection offers an interesting view on model adaptation to different data sources. The optimal parameter values are different: using pairwise function and a simple ReLU transformation for the first one; while using basically every option to counter overfitting on the second data source. Such different model structures arose from a combination of six different hyperparameters; a use case infeasible to test for a traditional parametric grid search.

The research of optimal computational methods for detecting AS soils will continue in the route of gathering and merging more available geospatial dataset.

References

1. Akusok, A., Björk, K., Miche, Y., Lendasse, A.: High-performance extreme learning machines: a complete toolbox for big data applications. IEEE Access **3**, 1011–1025 (2015)
2. Akusok, A., Espinosa Leal, L., Björk, K., Lendasse, A.: Scikit-elm: an extreme learning machine toolbox for dynamic and scalable learning. In: International Conference on Extreme Learning Machine, pp. 69–78. Springer (2019)
3. Beucher, A., Adhikari, K., Breuning-Madsen, H., Greve, M., Österholm, P., Fröjdö, S., Jensen, N., Greve, M.: Mapping potential acid sulfate soils in denmark using legacy data and lidar-based derivatives. Geoderma **308**, 363–372 (2017)
4. Beucher, A., Møller, A., Greve, M.: Artificial neural networks and decision tree classification for predicting soil drainage classes in denmark. Geoderma **352**, 351–359 (2019)
5. Beucher, A., Fröjdö, S., Österholm, P., Martinkauppi, A., Edén, P.: Fuzzy logic for acid sulfate soil mapping: application to the southern part of the finish coastal areas. Geoderma **226–227**, 21–30 (2014)
6. Beucher, A., Österholm, P., Martinkauppi, A., Edén, P., Fröjdö, S.: Artificial neural network for acid sulfate soil mapping: application to the sirppujoki river catchment area, south-western finland. J. Geochem. Explor. **125**, 46–55 (2013)

7. Boman, A., Sohlenius, G., Mattbäck, S., Becher, M., Liwata-Kenttälä, P., Öhrling, C., Auri, J., Edén, P.: Classification of finnish and swedish acid sulfate soil materials. In: Geophysical Research Abstracts, vol. 21 (2019)
8. Estevez Nuño, V.: Machine Learning methods for classification of Acid Sulfate soils in Virolahti. Master's thesis, Arcada University of Applied Sciences, Jan-Magnus Janssons plats 1, 00560 Helsinki, Finland (2020)
9. Estévez, V., Beucher, A., Mattbäck, S., Boman, A., Auri, J., Björk, K.M., Österholm, P.: Machine learning techniques for acid sulfate soil mapping in southeastern finland. Geoderma **406**, 115446 (2022)
10. Huang, G., Zhou, H., Ding, X., Zhang, R.: Extreme learning machine for regression and multiclass classification. IEEE Trans. Syst. Man Cybern. Part B (Cybern.) **42**(2), 513–529 (2012)
11. Huang, G.B., Zhu, Q.Y., Siew, C.K.: Extreme learning machine: theory and applications. Neurocomputing **70**(1), 489 – 501 (2006)

Online label distribution learning based on kernel extreme learning machine

Jintao Huang and Chi-Man Vong(✉)

Department of Computer and Information Science, University of Macau, Macau 999078, China
hjt.violler@connect.umac.mo, cmvong@um.edu.mo

Abstract. Label distribution learning(LDL) can reflect the significance of related labels in the sample compared to traditional label learning such as single-label learning (SLL) and multi-label learning (MLL). Label distribution learning problems are far more complex than SLL and MLL, as both the related labels and the significance of each class label described to each of the input samples are to be identified. Presently, more and more corresponding researches on label distribution learning are gradually emerging. However, almost all existing studies only consider batch learning when training the data without considering how to deal with streaming data. For this reason, a novel label distribution method is proposed in this paper to process online stream label distribution data, which is based on the idea of online learning and combined with an effective kernel extreme learning machine model. The proposed method can effectively deal with the stream data based on the designed sequential learning models. Finally, the experimental comparisons with the five commonly used label distribution learning models are conducted on 10 real label distribution benchmarks datasets, which further demonstrate the validity, feasibility, and stability of the method proposed in this paper.

Keywords: Label distribution learning · Online learning · Kernel extreme learning.

1 Introduction

In the field of machine learning, labeled ambiguity learning is a very hot topic. Presently, there are two more common machine learning paradigms to solve this problem: single-label learning (SLL) and multi-label learning (MLL) [1]. Unlike a single-labeled instance that can only be associated with one label, multi-label learning can associate multiple labels to an instance simultaneously which gives more meaning to each instance, so multi-label learning can solve practical problems more efficiently. However, the multi-label learning paradigm can only

K.-M. Björk (Ed.): ELM 2021, PALO 16, pp. 41–51, 2023.
https://doi.org/10.1007/978-3-031-21678-7_5

solve the problem of "whether the label describes the instance", and cannot solve the more complicated problem of "how much can the label describe the instance". In order to solve this problem, Geng et al. proposed a new paradigm namely label distribution learning (LDL) [2]. Differed from multi-label learning which outputs a label set, the output of label distribution learning is a label distribution. Each component in the distribution is a real number between 0 and 1, indicating the degree of description between the corresponding label and the instance of the component.

In the real world, the information implied by the relative weight of labels may be more common than many people think. A typical example is the emotional analysis of facial expressions. Facial expressions often convey a complex mixture of multiple basic emotions. Each basic emotion plays a different role in expression. The relative strength of all basic emotions naturally forms the emotional distribution of facial expressions. By treating emotions with the highest intensity or emotions with an intensity higher than the threshold as positive markers, SLL or MLL can be chosen as the solution to this problem. However, the price of this is that important information of different intensities of related emotions will be lost, which means that SLL or MLL framework alone cannot accurately predict the complex emotions of characteristics in real pictures. Therefore, the proposed label distribution learning paradigm (LDL) can give the distribution of label strength, so it can answer the question "How much does each label describe the instance?", and then more accurate prediction results can be obtained. That is to say, compared with traditional label learning, the LDL paradigm can provide richer instance description information, and the different relative importance of the labels assigned to the instance can better represent the label ambiguity.

After LDL was proposed, some scholars researched it and proposed many efficient algorithms. To solve the age prediction problem, Geng et al. converted a single-label set into a label distribution for training, and proposed the IIS-LDL algorithm [3], and then proposed a conditional probabilistic neural network algorithm based on neural networks [4]. And they also applied LDL in movie rating prediction [5] and aesthetic perception [6]. In addition, Xue et al. used LDL paradigm to try to predict personality tendency [7]. Zhang et al. proposed to use the LDL paradigm to count the crowd in video surveillance [8], and Sun et al. applied the LDL paradigm to identify abnormal behaviors in the video [9]. Wang et al. obtained the relationship between the label distribution model and the classifier model by in-depth analysis of the learning ability of the label distribution model [10]. And based on its theory [11], they replace KL divergence with absolute loss, use information entropy as sample weight, and propose a label distribution learning algorithm for classification. In terms of feature selection: To reduce the influence of noise redundant features, Xu et al. used latent semantics to train the label distribution model and performed feature selection at the same time. They proposed the LSE-LDL algorithm [12]. Ren et al. Determined by its corresponding part of the characteristics, by mining the specific characteristics corresponding to the mark to help model training, the LDLSF algorithm is proposed [13]. Huang et al. developed a label distribution feature selection algorithm using mutual information and label enhancement to remove some redundant or irrelevant features in multi-label data [14].

There are extensive state-of-the-art methods to solve the problem of label distribution learning. However, most of them are based on batch learning techniques, which involve collecting all data instances with prior knowledge and processing all the data to estimate parameters of the system simultaneously. In other words, batch learning technology needs to obtain all training data in advance without learning from streaming data. When need to learn from sequential stream data samples, it will pose a major limitation on the application of batch learning technology. Online learning, a series of machine learning technologies, is achieved by incrementally updating system parameters from sequentially arriving data using a one-way learning program [15,16]. In practical applications, the complexity of the naturally distributed data on the label is increasing, resulting in a very long training time, which poses a considerable challenge for the development of high-speed real-time online modules. Therefore, Online learning techniques are still great to be explored for label distribution learning. To do with this issue, this paper proposes an online label distribution learning method based on kernel extreme learning machines.

In order to effectively process the increased stream data, this paper is based on the idea of online learning, combined with the kernel extreme learning machine, through initializing the weights, proposes a method of continuous online learning and real-time update of the weights, which can effectively process the block of stream data. Compared with existing five state-of-the-art label distribution methods on ten benchmark label distribution datasets, the proposed method has achieved good performance on multiple data sets. This method is superior to the existing label distribution learning methods in performance and effectively process the increasing blocks of streaming data.

The rest of our paper is organized as follows. Brief reviews of label distribution learning and extreme machine learning are given in Sect. 2. Section 3 is the detailed information of proposed approach. The experimental comparisons and analysis are given in Sect. 4, and the Sect. 5 is the conclusion.

2 Preliminaries

2.1 Label Distribution Learning

The label distribution learning constructs a label distribution D_i for each instance x_i, and its components d_i^j represents the degree of description of the instance x_i by the component corresponding to the label y_j, $d_i^j \in [0,1], \sum_j d_i^j = 1$, that is, the instance can be completely described by the label in the label distribution. The description degree of all labels constitutes a data structure similar to probability distribution called label distribution, and the process of labeling sample learning is called label distribution learning [10,19,20]. As a new learning paradigm, single-label and multi-label can be regarded as special cases of label distribution learning under certain conditions, which makes the label distribution learning more generalized.

The existing label distribution learning algorithm strategies can be roughly divided into three categories [17]: problem transformation, algorithm adaption and special algorithm. Problem transformation is to transform the label distribution problem into a traditional learning paradigm that can be solved by existing learning paradigms such as PT-SVM and PTBayes algorithm. Algorithm adaption is to adapt certain machine learning algorithms to adapt to label distribution learning, such as the AA-BP algorithm and AA-kNN algorithm [3]. Specialized algorithms are designed to solve the problem of label distribution more directly included the SA-IIS algorithm. Designing a dedicated LDL algorithm mainly includes three parts: output model, objective function and optimization algorithm. These algorithms can effectively improve the prediction effect of label distribution learning.

The label distribution learning is to learn a mapping from a feature space to a set of label distributions composed of a finite set of labels. Let $X = [x_1, x_2, ..., x_n] \in R^{n \times t}$ represent the feature space where t is dimension of each instance. The finite label set is $Y = \{y_1, y_2, ..., y_L\}$, and the true label distribution is $D_i = [d_i^1, d_i^2, \ldots, d_i^L]$ corresponding to each instance x_i where L is a fixed number of labels in each label distribution. Although d_i^j does not mean the probability that y_j describes x_i correctly, it expresses that proportion of the degree which the label y_j described when a label set Y completely describes x_i. So LDL can also be expressed as a conditional probability density function learning from a training data set $T = \{(x_1, D_1), (x_2, D_2), ..., (x_n, D_n)\}$, which can obtain a similar distribution between conditional probability density function make the label distribution generated by this function for the unknown instance be as to the actual label distribution as possible.

2.2 Extreme Learning Machine

Neural network algorithms require more network parameter settings. When solving the optimal solution, a locally optimal solution may appear, and the global optimal solution cannot be obtained. Extreme learning machine (ELM) is a single layer feed-forward neural networks (Single Layer Feed-forward Neural networks, SLFNs) algorithm with high efficiency and good generalization [17,18]. In the ELM, only the number of hidden layer nodes of the network needs to be initially set, and the global optimal solution can be obtained without dynamically adjusting the input weights and biases of the network. The parameters of the hidden layer in the network structure of the algorithm are randomly selected. You only need to set the number of hidden layer network neurons, and finally use the least square method to find the output weight of the hidden layer. There is no need to iterate the network weight and bias in the process. Adjustment. Compared with the traditional neural network algorithm, the ELM algorithm has no iterative process, and has the characteristics of faster running speed and higher generalization performance.

Suppose N random samples $\{(X_i, Y_i) \,|\, i = 1, 2, ..., N\}$ where $X_i = [x_{i1}, x_{i2}, ..., x_{in}]^T$, and $Y_i = [y_{i1}, y_{i2}, ..., x = y_{im}]^T$. For L hidden layer nodes, the SLFNs is formally defined as:

$$\sum_{i=1}^{L} \beta_i g_i(X_j) = \sum_{i=1}^{L} \beta_i g(\omega_i \bullet X_j + b_j) = o_j. \tag{1}$$

where $\beta_i = [\beta_{i1}, \beta_{i2}, ..., \beta_{im}]^T$ is the weight of output, the $g(x)$ denotes the activation function, $\omega_i = [\omega_{i1}, \omega_{i2}, ..., \omega_{im}]^T$is the weight of input, and the b_i denotes the bias of i-th hidden layer neurons. The goal of a single hidden layer neural network is for the output error to be the smallest goal, which is formed as:

$$\sum_{j=1}^{N} ||o_j - y_j|| = 0. \tag{2}$$

According to formula (1) and formula (2), if given the $\beta_i, \omega_i$and b_i, the output of ELM classifier can be denoted as:

$$\sum_{i=1}^{L} \beta_i g(\omega_i \bullet X_j + b_j) = y_j. \tag{3}$$

And the formula(3) can be rewritten as:

$$H\beta = Y. \tag{4}$$

3 Proposed Approach

This paper uses the inherent high-speed characteristics of ELM and OS-ELM to develop an online sequential label distribution learning method for real-time streaming data applications. The key novelty of the proposed method is that there is no available online technology in the literature to perform real-time label distribution data. In single-label classification problems such as binary and multi-class classification, each input sample corresponds to a target label. Therefore, the classifier needs to identify a single target label corresponding to the input sample. In contrast, in multi-label classification, each input instance belongs to a subset of the target label. Therefore, the multi-label classifier needs to recognize the number and class of labels simultaneously. However, in label distribution learning, the model not only needs to recognize multiple category labels at the same time but also needs to analyze the degree of the description of each label for any instance, which makes the label distribution learning problem very complicated. From the aforementioned analysis, it can be seen that the existing label distribution models are all batch learning models without considering the advantages and characteristics of online learning when facing streaming data. To this end, this paper proposes an online label distribution learning model based on kernel extreme learning, which is called OELM-LDL. The detailed steps can be description as following:

1. Initialization and pre-processing the basic parameters of the kernel ELM network should be initialized, such as the number of hidden neurons and the kernel function. For the formula (4), the calculation result is easily to be affected by the random setting value. For this reason, the kernel matrix is introduced to solve this problem, and its formalization is as follows:

$$\Omega_{ELM} = HH^T : \Omega_{ELM(i,j)} = K(x_i, x_j). \tag{5}$$

where $K(x_i, x_j) = exp(-\gamma \mid\mid x_i - x_j \mid\mid)$. Consequently, the function of kernel extreme learning can be denoted as:

$$\begin{aligned} f(x) = &\quad h(x)H^T(\tfrac{E}{C} + HH^T)^{-1}Y \\ = &\begin{bmatrix} K(x, x_1) \\ \vdots \\ K(x_i, x_N) \end{bmatrix} (\tfrac{E}{C} + \Omega_{ELM})^{-1}Y \end{aligned} \tag{6}$$

where E is the identity matrix and C is regularization term.

2. Kernel ELM training (sequential learning) For the initialized training data N_0, the initial output weight β_0 can be calculated by $\beta_0 = \Omega_0 H_0^T Y$ where $\Omega_0 = (H_0^T H_0)^{-1} = (\Omega_{ELM})^{-1}$. Given each sequentially coming data/block of data D_{t+1} of N_{t+1} samples, the updated $\beta_{t+1}$of the network can be obtained by:

$$\Omega_{t+1} = \Omega_t - \frac{\Omega_t h_{t+1} h_{t+1}^T \Omega_t}{1 + h_{t+1}^T \Omega_t h_{t+1}} \tag{7}$$

$$\beta_{t+1} = \beta_t + \Omega_{t+1} h_{t+1}(Y_{t+1}^T - h_{t+1}^T \beta_t) \tag{8}$$

where $t = [0, 1, 2, ..., N - N_0 - 1]$. Specifically, the model only should be updated with one times when $N_{t+1} = 1$.

4 Experimentation

On 11 real data sets, 5 evaluation metrics are used to compare the algorithm in this paper with 5 existing label distribution learning algorithms.

4.1 Datasets and Settings

In order to verify the effectiveness of the proposed algorithm in this paper, 10 kinds of label distribution learning benchmark data sets are selected, as shown in Table 1. The Yeast data set is yeast gene data, SBU_3DFE is the face data set, and Movie is the crowd's scoring data set for movies. Human Genes is a dataset of human genes. All the detailed information of these 10 datasets can be found at http://cse.seu.edu.cn/people/xgeng/index.htm.

Table 1. 10 benchmark datasets for label distribution learning

Dataset	Instances	Features	Labels
Yeast-alpha	2465	24	18
Yeast-cdc	2465	24	15
Yeast-diau	2465	24	7
Yeast-dtt	2465	24	6
Yeast-elu	2465	24	6
Yeast-heat	2465	24	4
Yeast-spo	2465	24	4
Movie	7755	1869	5
SBU-3DFE	2500	243	6
HumanGene	30542	36	68

All the coding is run on matlab 2016b on windows10. In this paper, five commonly used evaluation metrics are selected for experimental comparison, namely Chebyshew, Clark, Kullback-Leibier, Cosine, and Intersection [2,18] to comprehensively evaluate the effectiveness of the label distribution learning algorithm. The first 4 metrics are distance measurement, and the smaller the value, the better. The latter two metrics are a measure of similarity, and the larger the value, the better.

In order to verify the effectiveness of the proposed algorithm, five mainstream comparison algorithms are selected for performance comparison with this research algorithm. These five algorithms are PT-Bayes, AA-KNN, AA-BP, SA-IIS, SA- BFGS. All the parameter settings of each comparison algorithm are selected as recommended by the original algorithm.

4.2 Results and Analysis

Since existing comparison algorithms cannot process streaming data, i.e., online learning cannot be performed. Therefore, the experimental part is mainly divided into two parts. The experiment adopts five-fold cross-validation. Each validation divides the data set into 80% train set and 20% test set, and finally calculates the average and standard deviation as the final result. For the first part, when compared with other algorithms, all training sets are directly verified. The evaluation results of 10 data sets are displayed in Tables 2, 3, 4, 5 and 6, and the best results are marked in bold.

From Tables 2, 3, 4, 5 and 6, it can be seen that the results of our proposed method perform better under five evaluation metrics. Among them, OELM-LDL achieves the best value on 4 Yeast data sets and the Movie data sets, and it performs well in the SBV_3DFE data set and the Human Gene data set. Additionally, OELM-LDL obtains the optimal rates on 10 datasets of 70%, 70%, 60%, 70%, and 60% in terms of the five metrics, i.e., Chebyshev, Canberra, KL,

Table 2. The results of Chebyshev(↓) on 10 benchmark datasets

Dataset	PT-Bayes	AA-BP	SA-IIS	SA-BFGS	OELM-LDL
Yeast-alpha	0.0995	0.0361	0.0171	**0.0132**	**0.0132**
Yeast-cdc	0.1091	0.0382	0.0200	0.0161	**0.0159**
Yeast-elu	0.1120	0.0381	0.0201	0.0161	**0.0150**
Yeast-diau	0.1576	0.0502	0.0411	**0.0360**	0.0371
Yeast-heat	0.1746	0.0533	0.0461	0.0420	**0.0415**
Yeast-spo	0.1759	0.0671	0.0611	**0.0582**	0.0591
Yeast-dtt	0.1813	0.0433	0.0431	0.0351	**0.0344**
Movie	0.2012	**0.0946**	0.1472	**0.1263**	0.1266
SBU-3DFE	0.1383	0.1426	0.1345	0.1042	**0.1015**
HumanGene	0.1954	0.0631	0.0542	**0.0531**	**0.0531**

Table 3. The results of Canberra(↓) on 10 benchmark datasets

Dataset	PT-Bayes	AA-BP	SA-IIS	SA-BFGS	OELM-LDL
Yeast-alpha	4.1932	2.4206	0.8617	0.6824	**0.6684**
Yeast-cdc	3.5418	1.7984	0.7833	**0.6477**	0.6886
Yeast-elu	3.2772	1.5980	0.7123	0.5844	**0.5752**
Yeast-diau	1.7084	0.5947	0.4785	**0.4270**	0.4468
Yeast-heat	1.4447	0.4616	0.4042	0.3678	**0.3589**
Yeast-spo	1.4439	0.5917	0.5391	0.5188	**0.5101**
Yeast-dtt	0.8710	0.2013	0.2023	**0.1695**	0.1705
Movie	1.5369	1.2286	1.1196	1.0645	**1.0144**
SBU-3DFE	0.9020	0.9832	0.8964	0.7285	**0.7036**
HumanGene	33.918	25.295	14.626	**14.516**	14.671

Table 4. The results of Kullback-Leibler(↓) on 10 benchmark datasets

Dataset	PT-Bayes	AA-BP	SA-IIS	SA-BFGS	OELM-LDL
Yeast-alpha	0.2773	0.0877	0.0084	**0.0063**	0.0073
Yeast-cdc	0.2832	0.0679	0.0095	0.0073	**0.0070**
Yeast-elu	0.2852	0.0682	0.0093	0.0062	**0.0060**
Yeast-diau	0.2697	0.0264	**0.0176**	0.0299	0.0303
Yeast-heat	0.2684	0.0269	0.1280	**0.0157**	**0.0157**
Yeast-spo	0.2789	0.0333	**0.0281**	0.2462	**0.0281**
Yeast-dtt	0.2265	0.0091	0.0080	**0.0060**	**0.0060**
Movie	0.7299	0.1666	0.1376	0.1196	**0.1073**
SBU-3DFE	0.0894	0.1032	0.0891	**0.0542**	0.0557
HumanGene	1.8059	0.5980	0.2464	0.2381	**0.2379**

Table 5. The results of Cosine(↑) on 10 benchmark datasets

Dataset	PT-Bayes	AA-BP	SA-IIS	SA-BFGS	OELM-LDL
Yeast-alpha	0.8457	0.9453	0.9910	**0.9940**	**0.9940**
Yeast-cdc	0.8503	0.9544	0.9901	**0.9930**	**0.9930**
Yeast-elu	0.8512	0.9573	0.9900	0.9902	**0.9910**
Yeast-diau	0.8637	0.9731	0.9831	0.9880	**0.9890**
Yeast-heat	0.8663	0.982	0.9841	**0.9877**	0.9871
Yeast-spo	0.8619	0.9653	0.9743	0.9772	**0.9801**
Yeast-dtt	0.8955	0.9910	0.9912	**0.9941**	0.9930
Movie	0.8459	0.9024	0.9011	0.9230	**0.9300**
SBU-3DFE	0.9170	0.9036	0.9201	**0.9403**	**0.9403**
HumanGene	0.4507	0.6869	0.8313	0.8332	**0.8321**

Table 6. The results of Intersection(↑) on 10 benchmark datasets

Dataset	PT-Bayes	AA-BP	SA-IIS	SA-BFGS	OELM-LDL
Yeast-alpha	0.7727	0.8743	0.9511	**0.9629**	0.9624
Yeast-cdc	0.7714	0.8866	0.9470	0.9572	**0.9590**
Yeast-elu	0.7729	0.8915	0.9497	**0.9585**	0.9580
Yeast-diau	0.7698	0.9178	0.9333	0.9401	**0.9474**
Yeast-heat	0.7702	0.9233	0.9336	0.9396	**0.9400**
Yeast-spo	0.7698	0.9025	0.9106	**0.9157**	0.9152
Yeast-dtt	0.7969	0.9493	0.9498	**0.9581**	**0.9581**
Movie	0.7227	0.7971	0.8033	0.8226	**0.8243**
SBU-3DFE	0.8383	0.8236	0.8397	**0.8702**	0.8613
HumanGene	0.4774	0.6367	0.7814	0.7834	**0.7855**

Cosine, and Intersection, respectively. And the analysis can be summarized as follows:

(1) According to the comprehensive performance of the algorithm, it can be shown that the algorithm in this paper is superior to other comparing algorithms in each evaluation index.
(2) Although the effect of OELM-LDL on SBV_3DFE data set and Human Gene data set is slightly lower than SA-BFGS algorithm, the overall performances are slightly better than SA-BFGS algorithm. On the SBU_3DFE data set, the feature difference between samples is small, so the performance is not good. On the Human Gene data set, because the data set is too large and easy to under-fit, the effect is slightly worse.
(3) The algorithm in the Yeast dataset and the Movie dataset performs better than other algorithms in all metrics.

5 Conclusions

Compared with single-label learning and multi-label learning, label distribution learning has more generalized performance in real applications. Existing label distribution learning methods use batch processing in the training process. This paper is oriented to streaming data and designs an online label distribution learning model based on the kernel extreme learning machine. It mainly includes model initialization and pre-processing, sequential learning mechanism. The method proposed in this paper can effectively solve the problem of online stream labeling and data distribution. And the experimental comparison, our proposed method outperforms the existing state-of-the-art label distribution learning methods in terms of both speed and performance.

References

1. Tsoumakas, G., Katakis.: Multi-label classification: an overview. Int. J. Data Warehouse. Min. **3**(3):1–13 (2007)
2. Xin, G.: Label distribution learning. IEEE Trans. Knowl. Data Eng. **28**(7), 1734–1748 (2016)
3. Geng, X., Zhou, Z.H.: Facial age estimation by learning from label distributions. In: Proceedings of 24th AAAI Conference on Artificial Intelligence (AAAI' 10), pp. 451–456 (2010)
4. Geng, X., Zhou, Z.H.: Facial age estimation by learning from label distributions. IEEE Trans. Pattern Anal. Mach. Intell. **35**(10), 2401–2412 (2013)
5. Geng, X., Hou, P.: Pre-release prediction of crowd opinion on movies by label distribution learning. In: Proceedings of the 24th International Conference on Artificial Intelligence, pp. 3511–3517 (2015)
6. Ren, Y., Geng, X.: Sense beauty by label distribution learning. In: Proceedings of the 26th International Joint Conference on Artificial Intelligence, pp. 2648–2654 (2017)
7. Xue, D., Hong, Z., Guo, S.: Personality recognition on social media with label distribution learning. IEEE Access **5**, 13478–13488 (2017)
8. Zhang, Z.X., Wang, M., Geng, X.: Crowd counting in public video surveillance by label distribution learning. Neurocomputing **166**, 151–163 (2015)
9. Sun, M., Zhang, D.P., Qian, L.Y.: Crowd abnormal behavior detection based on label distribution learning. In: Proceedings of the 4th International Conference on Intelligent Computation Technology and Automation, IEEE Press, pp. 345–348 (2015)
10. Wang, J., Geng, X.: Theoretical analysis of label distribution learning. In: Proceedings of the 33rd AAAI Conference on Artificial Intelligence, pp. 5256–5263 (2019)
11. Wang, J., Geng, X.: Classification with label distribution learning. In: Proceedings of the 28th International Joint Conference on Artificial Intelligence, pp. 3712–3718 (2019)
12. Xu, S.P., Shang, L., Shen, F.R.: Latent semantics encoding for label distribution learning. In: Proceedings of the 28th International Joint Conference on Artificial Intelligence, pp. 3982–3988 (2019)

13. Ren, T.T., Jia, X.Y., Li, W.W.: Label distribution learning with label-specific features. In: Proceedings of the 28th International Joint Conference on Artificial Intelligence, pp. 3318–3324 (2019)
14. Qian, W.B., Huang, J.T., Wang, Y.L., Shu, W.H.: Mutual information-based label distribution feature selection for multi-label learning. Knowl. Based Syst. **195**, 105684 (2020)
15. Venkatesan, R., Er, M.J., Dave, M., et al.: A novel online multi-label classifier for high-speed streaming data applications. Evolving Syst. **8**, 303–315 (2017)
16. Du, J., Vong, C.M.: Robust Online Multilabel learning under dynamic changes in data distribution with labels. IEEE Trans. Cybern. **50**(1), 374–385 (2020)
17. Geng, X., Xu, N.: Label distribution learning and label enhancement. Sci. sinica: Inf. **28**(5), 521–530 (2018)
18. Geng, X., Xu, N., Shao, R.F.: Label enhancement for label distribution learning. J. Comput. Res. Develop. **54**(6), 1171–1184 (2017)
19. Qian, W.B., Huang, J.T., Wang, Y.L., et al.: Label distribution feature selection for multi-label classification with rough set. Int. J. Approx. Reason. **128**, 32–55 (2021)
20. Jia, X.Y., Li, Z.C., Zheng, X., et al.: Label Distribution Learning with Label Correlations on Local Samples. IEEE Trans. Knowl. Data Eng. **33**(4), 1619–1631 (2021)

A New Energy Vehicle Thermal Runaway Data Processing Model Based on Machine Learning Algorithm

Liang Xiaoming[1], Fu Zhen[1], Liu Xiangchao[1(✉)], Peng Kai[1], Chang Wei[2], and Li Yuan[2]

[1] First Auto Work (Nanjing) Co. Ltd, Nanjing 211100, China
874062567@qq.com
[2] Shanghai Cloud Ready Technology Co. Ltd, Shanghai 200030, China

Abstract. By monitoring the battery and vehicle state data of electric vehicles in different states, we established the thermal runaway data model by using the streaming data based on time series as the data source. Firstly, we preprocessed the outliers, null values and numerical conversion for voltage, temperature, and other data. Secondly, we also carried out the multi-dimensional expansion for indicators that can effectively reflect thermal runaway, such as voltage, temperature, SOC and entropy. Besides, we obtained an additive model based on the characteristics of the indicators. Through the real thermal runaway test on the data of 45 normal vehicles, we calculated the thermal runaway score in the whole life cycle, and evaluated the effect according to the established thermal runaway threshold. The results show that the model and algorithm described in our paper can effectively carry out thermal runaway warning and realize false positives within the range of very low probability, which has important theoretical significance and application value.

Keywords: Thermal runaway processing · Flow data prediction · Machine learning algorithm · New energy vehicles

1 Introduction

For new energy battery vehicles, the most dangerous accident that power battery is prone to is thermal runaway [1]. Thermal runaway of battery refers to the phenomenon of spontaneous combustion caused by excessive junction temperature, which is a common electrothermal positive feedback in bipolar transistors. If the temperature rise of the battery reaches a certain threshold, the working state of the battery will be completely out of control, resulting in a straight rise in the internal temperature of the battery [2], or even combustion and explosion [3]. Therefore, it is very important to monitor the data indicators related to battery thermal runaway in real time and to give early warning to the phenomenon of battery thermal runaway.

An additive model consisting of six sub-models is established to predict the probability of thermal runaway in the future time axis. Different machine learning algorithms

K.-M. Björk (Ed.): ELM 2021, PALO 16, pp. 52–67, 2023.
https://doi.org/10.1007/978-3-031-21678-7_6

are used to build sub-models according to different influencing factors, and thermal runaway test and effect evaluation are carried out according to the data of 45 normal vehicles, so as to provide important basis and reference for real-time warning of thermal runaway phenomenon.

2 Data

2.1 Data Preparations

Battery data of electric vehicles include monitoring data of electric vehicles, which will be generated in real time in different vehicle states of electric vehicles—driving and charging. Real-time monitoring data will be collected every 10 s. The battery monitoring data includes the battery itself data and vehicle status data in the normal use process, totaling more than 200 data variables. In this paper, the data of 45 normal vehicles of the same model and the data of 21 thermal runaway vehicles (one of which is the thermal runaway data of FAW, and the other 20 are the thermal runaway data of other vehicles) are provided as the data source for thermal runaway modeling and prediction.

The data used are all flow data based on time series, including current, voltage, temperature, pressure difference, pressure rise, temperature difference, temperature rise, entropy and so on related to machine learning. The relevant data content is shown in Table 1 (Table 2).

2.2 Data Exploration

1. Total data: 45 vehicles on board and 21 vehicles with thermal runaway (one of which is FAW car's data).
2. Data fields: Data fields meet the requirements. (It means that all feature fields required by capacity extraction and modeling process exist in the original and data.)
3. Data accuracy: Except for the low accuracy of SOC, the accuracy of other data fields meets the requirements.

   Specific precision requirements are as follows:

4. Data quality:

   Good quality of time data, few repeat time data, and basically stable 10-s interval (Tables 3 and 4).

2.3 Data Preprocessing

The data preprocessing mainly includes some outliers, vacancy and numerical conversion in the original data fields, as well as extraction and transformation of single voltage and temperature measuring point temperature, etc., as follows:

Table 1. Data fields.

Serial number	The original field	Description
1	gb_collect_time	Time stamp, normal collection frequency is 10 s
2	gb_charging_status	The operating mode of the vehicle, including driving, charging, etc
3	gb_soc	The state of charge of a battery, which represents the current remaining capacity of the battery
4	gb_mileage	Accumulated mileage of vehicle travel
5	gb_pack_current	The charge and discharge current of the battery pack
6	gb_pack_volt	Total battery pack voltage
7	gb_cell_max_volt	Maximum value of single voltage
8	gb_cell_min_volt	Minimum value of cell voltage
9	gb_max_temp	The maximum temperature of the temperature measuring point
10	gb_min_temp	The lowest temperature at the temperature measuring point
11	gb_cell_volt_group_value	1 ~ N battery voltage values were collected
12	gb_cell_temp_group_value	The temperature of 1 ~ m temperature detection points collected

1. For data de-duplication, as long as there are multiple pieces (or each piece) of completely repeated data in each car, only one of them will be retained and the remaining pieces of data will be deleted.
2. Time stamp outlier processing. Set value ranges for year, month, day, hour, minute and second. For example, the value range of month is 1–12, and the value range of day is 1–31, etc. Data beyond the range will be deleted entirely.
3. Extraction of monomer voltage and temperature measuring point temperature. In the original data, the monomer voltage and temperature measuring point temperature are stored in a certain string list, which will be particularly troublesome and time-consuming in practical application. Therefore, each monomer voltage and temperature measuring point temperature are extracted as a separate column of data. Finally, all the data formats of individual voltage and temperature measurement point temperature are converted to INT16 data type, which will take up the minimum memory space.
4. Temperature correction. The temperature in the original data is not the real temperature, so 40 should be subtracted to get the real temperature data.
5. Correction of charge/discharge current. The charge/discharge current in the original data is not the real charge/discharge current, which needs to be divided by 10 and subtracted by 1000 to get the real charge/discharge current data.

Table 2. Field calculation logic.

Serial number	Field	Calculation logic
1	Differential pressure	Maximum cell voltage minus minimum cell voltage
2	Pressure increase quantity	The result of subtracting all cell voltages of the previous data from all cell voltages of the current data. The number of voltages is 96
3	Voltage entropy	Entropy calculated from all cell voltages
4	Temperature difference	The temperature value of the highest temperature measuring point minus the temperature value of the lowest temperature measuring point
5	Temperature rise of the amount	The result of subtracted all temperature points of the previous data from the temperature of all temperature points of the current data. The number of temperature rises is 48
6	Current differential	The difference between the electric current and the current at the previous time
7	Temperature entropy	Entropy values calculated at all temperature measurement points
8	Time difference	Current time minus the time of the previous data (in units per second)
9	Number of historical "overcharges"	The charging behavior of SOC > 90 at the end of each charge was counted

Table 3. Precision requirement.

Field	SOC	Current	Monomer voltage	Time span
Accuracy requirement (minimum)	One decimal place	One decimal place	mv	$<=10$ s

Table 4. Data quality.

Time span	< 10 s	10 s	> 10 s
Quality assessment	Very good	Good	Poor

6. The total voltage of the battery pack is corrected. The data of the total voltage in the original data is not the real total voltage, which needs to be divided by 10 to get the real total voltage data.
7. In the follow-up steps such as flux extraction, targeted data processing will also be carried out on the accumulated mileage and other data.

Here are the new data preprocessing tasks for code optimization:

8. For the data in which the monomer voltage and the temperature of the temperature measuring point are all 0, the final score is set to 0 by default.
9. For the data in which the minimum temperature of the temperature measuring point is 0 and the maximum temperature is less than 5, the final score is 0 by default.
10. For the data in which the maximum value of the monomer voltage is equal to the minimum value of the monomer voltage is equal to 3.650, the default final score is 0.
11. For data with single voltage less than 0 or greater than 4.4, the final score is set to 0 by default.
12. If the temperature of the temperature measuring point is less than 0 or more than 200, the final score will be 0 by default.

3 Feature Extraction

For the above data, through the preliminary analysis, we believe that voltage, temperature, current, SOC, entropy and other indicators can more effectively reflect the thermal runaway. Based on Beijing new energy rich data dimensions at the same time, for each indicator, the multiple dimensions of development, such as the voltage and temperature, we focus on the value of itself, but also on changes in quantity, for multiple monomers by information entropy to measure its consistency, suitable for high precision data fitting to determine the distribution of abnormal values and so on. During the whole modeling process, we mainly extracted the following features in Table 5.

4 Model Building

In order to deeply explore the influence of each index on thermal runaway and ensure the good extensibility of the model, we combined the characteristics of each index to establish the following additive model,

$$Score_1 = \beta_1 * V_d + \beta_2 * T_d + \beta_3 * E_a + \beta_4 * S_{OC} \tag{1}$$

$$Score_2 = \text{Full cut} - \text{off differential pressure hazard factor} \tag{2}$$

where V_d is voltage density, T_d is temperature density, C_d is current distribution, E_a is entropy anomaly, S_{OC} is SOC & over-charge risk, and C is constant term (mainly battery architecture, etc.).

Table 5. Feature description.

Feature	Description
CellVoltage	The voltage of each cell
CellVoltageDiff	The difference between the highest and lowest cell voltages
CellVoltageDelta	The difference between the next moment and the last moment of the voltage of each cell
InnerVoltage	The total voltage inside the power battery
HighVoltage	The highest single voltage of the power battery
LowVoltage	The lowest single voltage of the power battery
CellTemp	Temperature of each cell
CellTempDiff	There is no difference between the temperature at the high and the lowest temperature point
CellTempDelta	The difference between the temperature of each cell at the next time and the last time
HighTemp	The highest single temperature of the power battery
LowTemp	The lowest single temperature of the power battery
Current	Power battery charge/discharge Current
CurrentDelta	The difference between the charge/discharge current of a power battery at one moment and the last moment
VoltEntropy	The information entropy obtained from calculation of monomer voltage
Tempentropy	The entropy of information calculated by the temperature of the temperature measuring point
Point	A Point in the two-dimensional coordinates of Entropy_colt and Entropy_temp
Soc0	The SOC at the start of charging
Soc1	The SOC at the end of the charge
Soc_delta	End SOC minus Start SOC represents the amount of change in SOC during charging
OverChargeTimes	The total number of overshots in history
TimeDiff	The difference between the timestamp of the current received data and the previous data timestamp

This model consists of two parts, Score1 and Score2. The Score1 model contains six parts: voltage model based on probability density, temperature model based on probability density, abnormal current identification model based on distribution, consistency determination model based on information entropy, overcharge risk assessment model based on SOC, the influence of other factors on thermal runaway evaluated by constant term. The Score2 model is full cut-off differential pressure hazard factor model. As for the parameters of the additive model, we choose better parameters based on the range provided by experts to identify more abnormal cars as the goal. We will briefly describe each of the sub-models below.

In addition, considering the complex situation of different vehicle types, we suppose that each vehicle type is independent of each other. We build thermal runaway models for them respectively. The Score1 model contains the following A to E parts:

4.1 Logistic Model of Voltage and Temperature Based on Probability Density

For the two indicators of voltage and temperature, we derived many other features, the specific items are as follows:

(1) CellVoltage
(2) CellVoltageDiff
(3) CellVoltageDelta
(4) InnerVoltage
(5) HIghVoltage
(6) LowVoltage
(7) CellTemp
(8) CellTempDiff
(9) CellTempDelta
(10) HIghTemp
(11) LowTemp

Where features (1)–(3) indicate the monomer voltage value, the monomer voltage difference, the monomer voltage difference at the next moment compared to the previous moment, (4)–(6) indicate the total voltage value, the highest voltage value and the lowest voltage value, respectively. (7)–(11) are similar to (1)–(6) and indicate the monomer temperature conditions.

Finally, we extracted 196 voltage-related features and 99 temperature-related features. We need to find the vehicle numbers with thermal runaway faults by using probability density and logistic regression prediction models for multiple samples of the feature data. Therefore data training and model building was carried out on the above 11 data features.

For the above features, we extracted features from the data of all normal cars and took them as positive samples. For the thermal runaway vehicle, we extracted features from a period of time before thermal runaway, as a negative sample.

We divided the corresponding feature interval options for each feature, and then calculated the probability function of each feature. Since outliers are a small part of the overall data, we calculated all probability densities of all features. After that, In order to distinguish between abnormal or normal feature taking values, after many experiments and selections, we finally selected the top 10 feature probability density values (the lower the ranking is, the higher the ranking is) as the training data. The probability density calculation formula is as follows,

$$density = \frac{D_i}{D_t} \tag{1}$$

where D_i is the amount of data within a certain interval, D_t is the total amount of data.

Now we obtain the top 10 characteristic probability density values of positive samples and negative samples. Then we apply the Logistic model to train these numbers. Finally, we build the logistic model of voltage temperature on probability density.

After using the offline data for the above feature training, the prediction of the real-time online data also needs to be carried out. We find the interval corresponding to the value of each feature, for each interval to calculate the corresponding probability density value, on this basis for real-time data prediction, because in the training process has been for the normal vehicle number labeled 0, abnormal thermal runaway vehicle number labeled 1, so on this basis directly for the prediction of recognition can be.

4.2 Entropy-Based Monomer Consistency Determination Model

For monomer consistency judgment, we use the information entropy as the basis, the monomer voltage and temperature measuring point of information entropy combined into a single point in the 2 d coordinates [4], so that each of the data will be a voltage temperature point of information entropy, and then the normal car to get to the point of data being entered OneClassSvm algorithm model, training model of a contain so normal data point [5]. When a new data point is passed in, the model determines whether it is a normal point (BBB 0 0) or an abnormal point (<0). The model effect drawing is shown in Fig. 1. Where the x-axis represents the entropy of the voltage characteristic and the y-axis represents the entropy of the temperature characteristic.

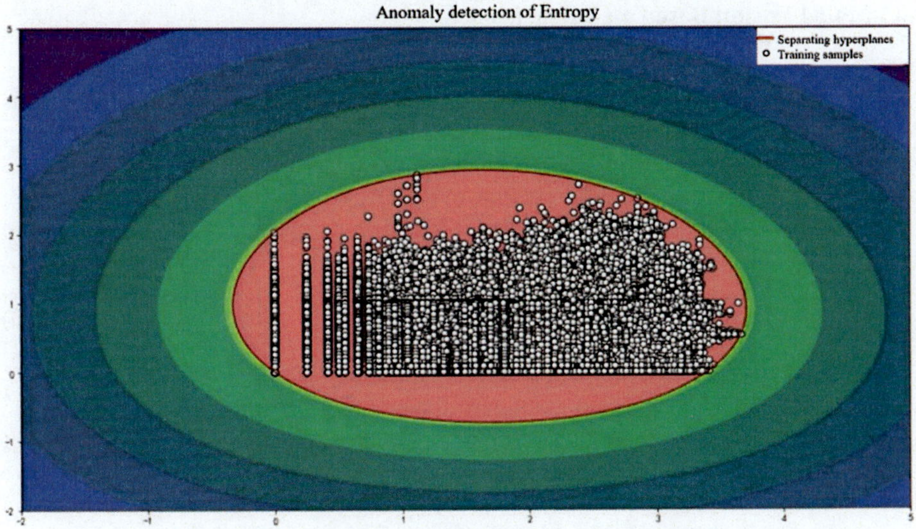

Fig. 1. Rendering of entropy-based monomer consistency determination model

The entropy calculation formula is as follows:

$$Entropy = -\sum_{i=1}^{n} p_i \log p_i \tag{2}$$

All normal points are divided inside the hyperplane. Each time a piece of data comes in, the model will calculate whether the point composed of voltage and temperature information entropy is in the hyperplane. If it is, a number greater than 0 will be returned, representing the normal point. Otherwise, it will return a number less than 0, indicating the abnormal point. In addition, if the value less than 0 is farther from 0, it means that it is farther from the segmentation hyperplane, that is, the degree of anomaly is relatively large.

4.3 Risk Factor Assessment Based on Real-Time SOC and Historical Charging Behavior

The first is the current SOC, we will only process the data whose current SOC is greater than or equal to 90, and directly return 0 if it is not. The second is the number of overcharging times in history. Depending on the current value of the SOC between 90 and 100, we will assign a different risk factor. And then according to the different history of the number of overshoot to different levels of battery loss. Finally, we calculated the final coefficient of the thermal runway combined with these two indexes.

4.4 Model Fusion Based on Single Model

We give each single model a model coefficient, the specific coefficient is obtained by cross validation (that is, each coefficient is given a certain range, in this case, 100 coefficients between 0 and 1 are taken), and the final coefficient is composed. Here, we also consider the influence of different battery pack structures and materials (ternary lithium, lithium iron phosphate, etc.). If it is a different type of battery, we need to consider the influence of the whole constant term. The final model is as follows (where β_1, β_2, β_3 and β_4 are the corresponding coefficients of the model):

$$Score_1 = \beta_1 * V_d + \beta_2 * T_d + \beta_3 * E_a + \beta_4 * S_{OC} \tag{3}$$

The Score2 model contains is described as follows:

We conducted multi-dimensional analysis on vehicles with thermal runaway and found that the cut-off pressure difference of full charge may be the potential factor of thermal runaway (cut-off pressure difference of full charge refers to the cut-off pressure difference of SOC = 100 at the end of charging). The following figure shows the variation of cut-off pressure difference of each full charge of vehicles with thermal runaway:

It can be clearly found that before thermal runaway (red dots on the way) occurs, there are two situations in which the pressure differential rises greatly. When the pressure differential rises greatly in the second full charge, thermal runaway occurs. In addition, the first rise in differential pressure is not an accidental rise or a statistical problem,

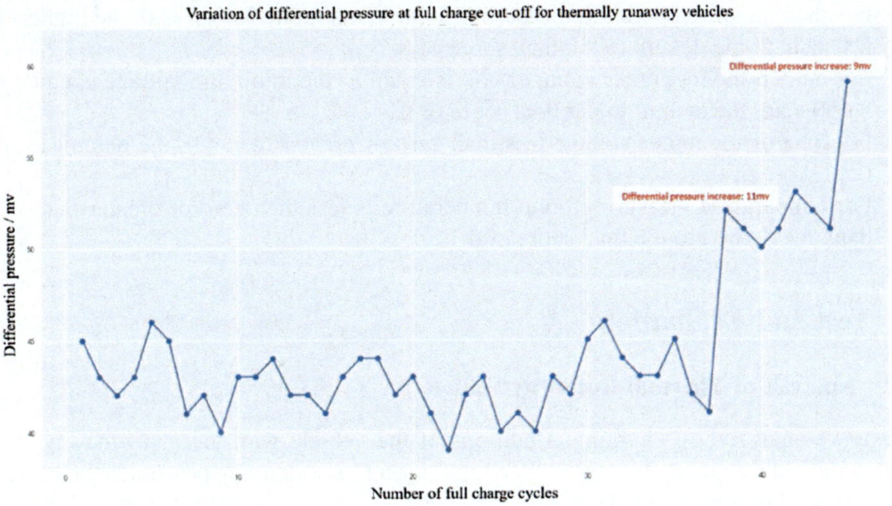

Fig. 2. Variation diagram of the cutoff pressure difference of the vehicle with full charge for thermal runaway

because after the first rise, the subsequent differential pressure remains about the same as after the first rise (Fig. 2).

As a result, we put forward 'Full cut-off differential pressure hazard factor' as Score2 to build a complete model.

5 Model Solution

5.1 Data Test

We conducted thermal runaway test on the data of 45 normal vehicles, and found that there were certain alarm situations (these alarms were actually false alarms). Combined with the test results of 45 vehicles, the false alarms were finally divided into three situations:

1. Data exist in a large number of continuous voltage and temperature is 0 (at most).
2. Data exists in the continuous part minimum temperature is 0, the highest temperature is less than 5.
3. Data in the presence of discrete data monomer voltage too high and too low.

5.2 Code Optimization

In view of the specific situation of these false positives, specific optimization measures are taken for the code, as follows:

1. For all data in the monomer voltage and temperature measuring point temperature is 0, the default to give to the final score of 0.

2. For the data of temperature measuring point temperature lowest is 0, the highest is less than 5, the default to the final score of 0.
3. For data in the monomer voltage peak is equal to the minimum voltage is equal to 3.650 data, the default to the final score of 0.
4. For data of monomer voltage less than zero or greater than 4.4, the default to the final score of 0.
5. For temperature measuring point temperature is less than zero or greater than 200 data, the default to the final score of 0.

6 Test and Evaluation

6.1 Analysis of Thermal Runaway Vehicle

Next, we analyzed all charging conditions of the vehicle with thermal runaway. The following is the change of pressure difference during each charging process of the vehicle with thermal runaway (see the part in the data set):

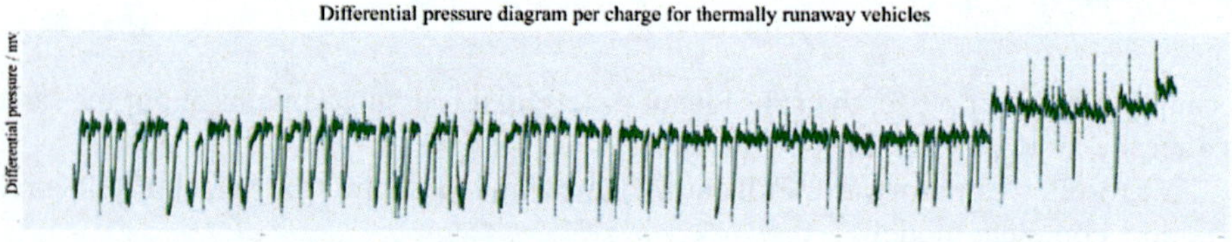

Fig. 3. Differential pressure diagram of thermal runaway vehicle per charge

It can be found from the Fig. 3 that there are two sudden increases in pressure differential before thermal runaway occurs, in which thermal runaway occurs in the second sudden increase in pressure. These two sudden increases correspond to the two sharp increases in pressure differential in the figure above.

6.2 Analysis of Normal Vehicle

Through the analysis of 45 normal vehicles with long mileage at the same latitude, we found that there was indeed a difference between them and the vehicles with thermal control. The following figure shows the distribution scatter diagram of the maximum pressure difference of full charge cutoff for 45 normal vehicles, as follows:

We found that the maximum differential pressure rise of the 45 normal cars with full charge cutoff did not reach that of the car with thermal runaway (10.7 for the first time and 8.0 for the second time) (Fig. 4).

The following is the pressure difference changes of 45 normal cars during all charging processes. For reasons of space, here are the pictures of several cars that can include all conditions (see the part in the data set):

1. Normal car 14 - constant drop in differential pressure

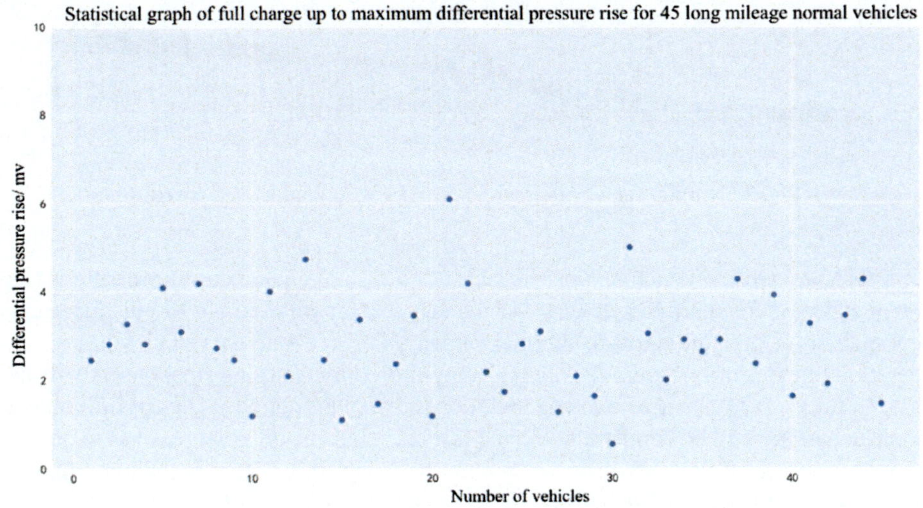

Fig. 4. Statistical chart of maximum differential pressure rise up to full charge of 45 normal vehicles with long mileage

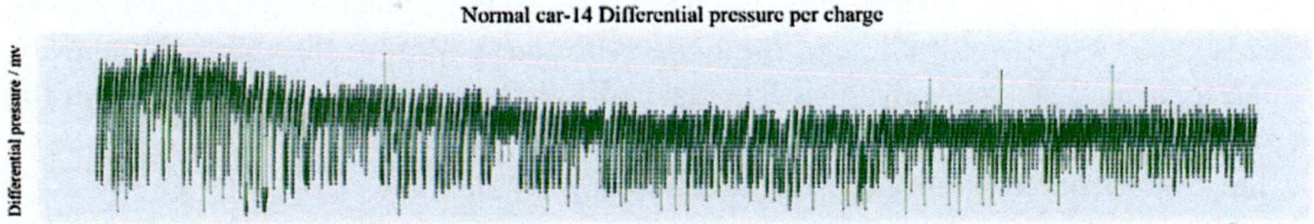

Fig. 5. Normal car 14 - Pressure difference diagram for each charge

2. Persistent pressure difference level of normal car 28

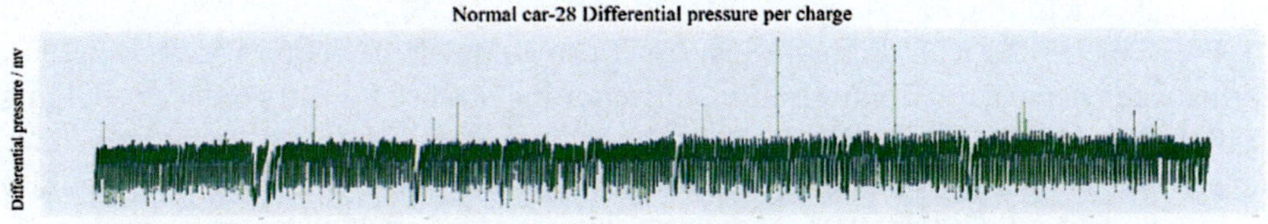

Fig. 6. Normal car 28 - Pressure difference diagram for each charge

3. Normal car 31 - pressure differential continues to rise

As shown in the Figs. 5, 6 and 7, in the process of normal car charging, the change of pressure difference is persistent, rather than the direct sudden rise of runaway vehicle.

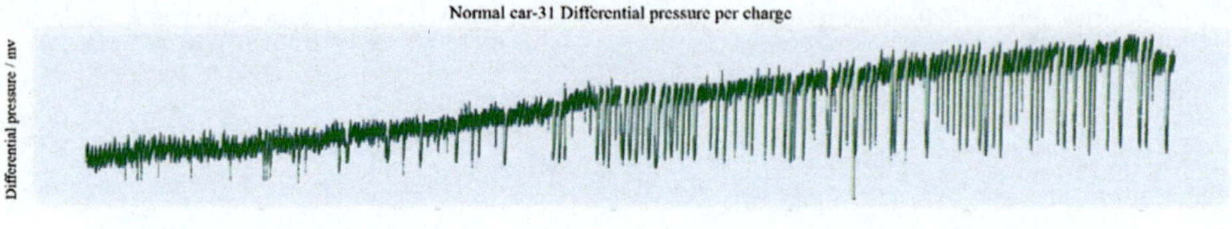

Fig. 7. Normal car 31 - Pressure difference diagram for each charge

Therefore, we can conclude that the sudden change of charge cutoff pressure is the potential factor for thermal runaway. At the same time, we are able to put forward a new method to give an alarm to thermal runway. For new energy vehicles in normal operation, the cut-off pressure difference is monitored by charging (most users will be charged fully), and the cut-off voltage mutation threshold is defined. The cut-off voltage of each charging will be compared with the last charging cut-off voltage, so as to obtain the real-time cut-off pressure difference mutation value. If it is greater than or equal to the set mutation threshold, an alarm will be given at high risk of thermal runaway.

6.3 Testing Results

The thermal runaway score threshold we set is 0.43, so when the thermal runaway score exceeds 0.43, it will return 1, and the code will alarm. Before the thermal runaway, the thermal runaway score of this car is basically maintained below 0.2, but when the thermal runaway occurs, the score has exceeded 0.43 and reached about 1.

The test results of 45 test vehicles are shown in Table 6.

6.4 Performance Evaluation

After optimizing the code, we tested 45 test data again. In addition, we tested the data of the most important vehicle with thermal runaway. The specific results are as follows:

The test results of thermal runaway vehicle data are shown in Table 7.

The final score of thermal runaway of a thermal runaway vehicle in its whole life cycle is shown in Fig. 8.

For vehicle data with thermal runaway, it can complete 42 min early warning. For normal vehicle data, after optimization, misreporting has been greatly reduced. Judging from the test results of 45 vehicles, if false positives are taken into account by vehicle dimension, there are 4 vehicles in the 45 test data with false positives, and the false positives rate is all within 0.01%. If the data dimension of all 45 vehicles is considered, the false positives proportion will be several orders of magnitude less. Therefore, for the thermal runaway algorithm this time, it has realized the real warning with high efficiency and the false positives with very low probability, which has become an applicable and excellent thermal runaway warning algorithm.

Table 6. Test results of 45 test vehicles.

Vehicle number	Misreporting type	Misreporting proportion (%)
00000	N/A	0
00001	(3)	0.008
00002	N/A	0
00003	(3)	0.003
00004	N/A	0
00005	N/A	0
00006	N/A	0
00007	N/A	0
00008	N/A	0
00009	N/A	0
00010	N/A	0
00011	N/A	0
00012	N/A	0
00013	N/A	0
00014	N/A	0
00015	N/A	0
00016	N/A	0
00017	N/A	0
00018	N/A	0
00019	N/A	0
00020	N/A	0
00021	N/A	0
00022	N/A	0
00023	N/A	0
00024	N/A	0
00025	N/A	0
00026	N/A	0
00027	N/A	0
00028	N/A	0
00029	N/A	0
00030	N/A	0
00031	N/A	0

(continued)

Table 6. (*continued*)

Vehicle number	Misreporting type	Misreporting proportion (%)
00032	N/A	0
00033	N/A	0
00034	(3)	0.002
00035	N/A	0
00036	N/A	0
00037	(3)	0.004
00038	N/A	0
00039	N/A	0
00040	N/A	0
00041	N/A	0
00042	N/A	0
00043	N/A	0
00044	N/A	0

Table 7. Thermal runaway vehicle data test results.

Vehicle number	Misreporting type	Misreporting proportion	Warning time	Thermal runaway time	Early warning advance quantity
Thermal runaway 001	N/A	0	2020/10/11 14:48	2020/10/11 15:30	42 min
Thermal runaway 002	N/A	0	2020/8/24 10:38	2020/8/24 10:59	21 min
Thermal runaway 003	N/A	0	2020/9/02 15:27	2020/9/02 15:57	30 min
Thermal runaway 004	N/A	0	2020/10/11 14:48	2020/10/11 15:30	42 min
Thermal runaway 005	N/A	0	2020/12/21 16:03	2021/1/04 16:26	14 days and 23 min
Thermal runaway 006	N/A	0	2021/3/15 9:28	2021/3/15 9:27	−0.5min

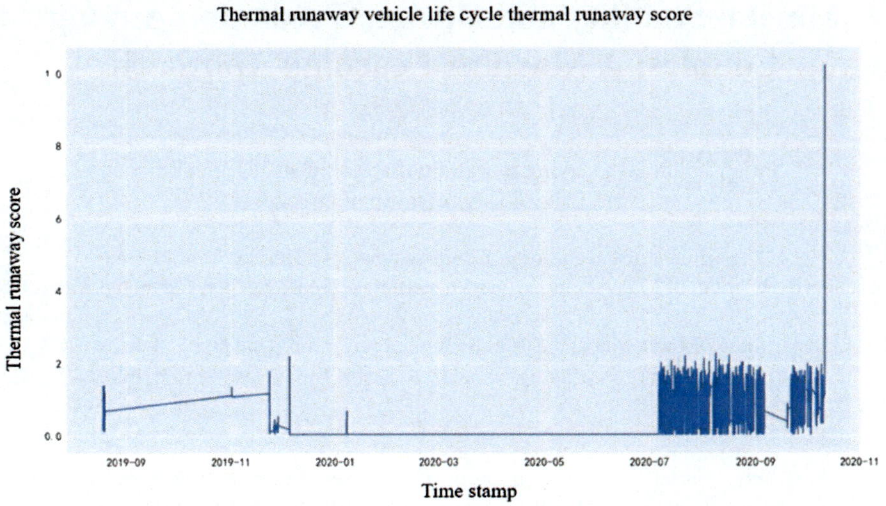

Fig. 8. Thermal runaway vehicle life cycle thermal runaway score variation trend chart.

References

1. Smith B.: Chevrolet Volt battery incident overview report [R]. US Department of Transportation. National Highway Traffic Safety Administration 1 (2012)
2. Citarella, M., et al.: IEEE Transportation Electrification Conference (ITEC-India). Bengaluru, India **2019**, 1–4 (2019). https://doi.org/10.1109/ITEC-India48457.2019.ITECINDIA2019-274
3. Qingsong, W., Ping, P., Xuejuan, Z., et al.: Thermal runaway caused fire and explosion of lithium ion battery[J]. J. Power Sources **208**, 210–224 (2012)
4. Zhang, G., Li D., Liu P., Zhang, Z.: Multi-dimension fault diagnosis of battery system in electric vehicles based on real-world thermal runaway vehicle data. In: 2019 IEEE sustainable power and energy conference (iSPEC), pp. 2830–2835, Beijing, China, (2019). https://doi.org/10.1109/iSPEC48194.2019.8975122
5. Ostanek, J., Li, W., Mukherjee, P., Crompton, K., Hacker, C.: Simulating onset and evolution of thermal runaway In Li-ion cellsusing a coupled thermal and venting model. Appl. Energy 15 June 2020 (2020)

Check for updates

Marital Stability and Divorce Prediction Among Couples: A Machine Learning Approach

Sadeq Fallahtafti[1], Alireza Fallahtafti[2](✉), Gary R. Weckman[3], and Hamide Mohammadinasab[4]

[1] Department of Psychology, Tarbiat Modares University, Tehran, Iran
[2] Weatherhead School of Management, Case Western Reserve University, Cleveland, Ohio, USA
axf466@case.edu
[3] Department of Industrial and Systems Engineering, Ohio University, Athens, USA
[4] Department of Psychology, Shahid Beheshti University, Tajrish, Iran

Abstract. Marital dissolution usually affects one's health destructively. This paper aims to examine significant factors in predicting marital stability among Iranian couples with high accuracy using machine learning algorithms. Machine learning algorithms can investigate significant features in marital stability, identify the most important features in such complicated information with multivariate data, and predict marital stability with high accuracy. Firstly, Young Schema Questionnaire–Short Form 2, Persian Divorce Prediction Scale, and Gottman Emotional Separation Scale are completed by participants. Then, two feature selection techniques, namely Univariate and Relief, are used. Afterward, four machine learning practices, including support vector machine (SVM), logistic regression (LR), decision tree (DT), and random forest (RF), are trained and tested on the set of full features alongside the selected features for their ability to predict marital stability with the highest possible accuracy. Several metrics are employed to evaluate the performance of multi-class classification predictors. After hyperparameter tuning, SVM and RF training algorithms achieved more than 89% accuracy on the created dataset. The findings can be utilized both by couples therapists and couples vulnerable to marital dissolution.

Keywords: Machine learning application · Supervised learning · Artificial intelligence · Marital Stability · Divorce prediction

1 Introduction

Although divorce can be a better choice [1] in some cases, numerous studies have found that the quality of close relationships between couples has a strong correlation with the physical and mental health of couples [2–4], their economic status [5], their children's well-being [6, 7], and even the health of individuals who were not yet born at the time of the original divorce [8]. Despite the benefits of a stable and happy marriage, statistics show that the number of stable marriages is declining, and more and more marriages are ending in two forms of marital dissolution, namely divorce or emotional separation. Divorce rates have been recently proliferating in many countries around the world.

K.-M. Björk (Ed.): ELM 2021, PALO 16, pp. 68–83, 2023.
https://doi.org/10.1007/978-3-031-21678-7_7

In Iran, the ratio of marriage to divorce was 6.1, 4.4, and 2.8 in 2011, 2014, and 2019, respectively [9]—an alarming trend. Studies show an analogous trend in other countries. For instance, in the United States, more than one million marriages end up in divorce each year [10]. Besides, within the COVID-19 crisis period, the struggle between couples can increase [11]—due to several possible reasons like lower sexual intimacy, death of loved ones, unemployment, and mental illness—and subsequently, divorce rates will probably increase. Divorce is not the only opposite side of a stable marriage. Emotional separation is another eminent form of failure in marriage [12], which is often regarded as the first stage in the process of divorce and is regarded as the sign of a deteriorating marriage saturated by loneliness [13]. Emotional separation is a stage in a marital relationship in which love, affection, and compassion are replaced with anger, despair, and disgust, which causes the couple to feel pessimistic about getting back on track. Numerous studies have shown that emotional separation is a significant predictor of divorce, has significant destructive effects on the couple's health, and can lead to depression, anxiety, and low self-efficacy [14]. It has been found to have detrimental effects on children [15] as well. Karlsen, Futris, and Richardson [16] found that couples who suffered from relationship uncertainty, one of the symptoms of emotional separation, reported more damaging behaviors that could lead to more chance of marital dissolution. Although divorce rates have been quickly increasing in the last several years in Iran, even in the cases where it seems the best decision for the couple, divorce is not encouraged as a common choice because, in many cases, it can lead people to feel stigmatized by it and therefore avoid it [17]. As a result, regarding the fact that many couples are unhappy in their marriages and feel helpless to figure out a solution but still prefer not to get divorced, it is reasonable to consider emotionally separated couples as a group of married people distinct from still happily married ones to understand this phenomenon in a better way.

Numerous tools such as interviews and questionnaires are used to assess the quality of marital relationships regarding marital stability and an accurate prediction of separation. However, analyzing such complicated information with multivariate data, including cultural factors, sounds quite tricky. To address this issue and develop a tool to predict whether a marriage will end in dissolution, we can implement machine learning (ML) algorithms as a computer-human analysis approach aimed at achieving high accuracy outcomes, thanks to its better functionality in investigative data analysis. Despite abundant studies in human behavior and ML applications, the body of literature utilizing computers in human behavior—especially ML in couples' behavior—is relatively narrow. Motivated by a case study, this paper examines high important features in marital stability. It puts participants into three distinct categories of happily married, emotionally separated, and divorced using ML algorithms. With that aim, we first apply feature selection methods such as Univariate and Relief to extract essential attributes from the dataset. Afterward, four machine learning practices, including support vector machine (SVM), logistic regression (LR), decision tree (DT), and random forest (RF), are employed and tested on the set of full features alongside the aforementioned selected features. The performance of multi-class classifiers is evaluated through various metrics. The results of this study offer a promising approach to predicting marital stability, which can be a significant source of guidance in marriage counseling and couples therapy settings.

2 Background

Studies have found different kinds of socio-economic [18–20], ethnical [21, 22], psychological [23–25], and interpersonal [26, 27] factors accounting for marital stability. In addition, some studies investigated the educational intervention's influence on reducing divorce petitions [28]. Numerous papers have studied the impact of early maladaptive schemas on marital satisfaction and divorce among Iranian couples [29, 30]. As defined by Jeffery Young [31], schemas are structures and frameworks as well as models of cognitive content which are developed in the early stages of one's life, leading to one's general view of life in an adaptive, maladaptive way. They are considered cognitive plans, guiding problem-solving skills and the interpretation of information. Schemas can shape spouses' understanding and interpretation of each other's behaviors and, therefore, significantly impact their marital satisfaction and their chance of separation.

Due to the successful application of artificial intelligence in human behavior and social issues, there is a large body of work on this topic in literature. For instance, researchers have used machine learning in the prediction of academic performance of college students with internet usage behavior [32, 33], human-defined artificial intelligence behavior patterns in games [34], human activity recognition [35], and understanding emotions in textual dialogues [36]. However, to the best of the authors' knowledge, few papers have studied ML techniques in the domain of couples' marital stability/dissolution. Guillén et al. [37] presented an ML application that predicted the type of ending of marital status—either in a friendly agreement or being taken to court—when marriage was in a dissolution process. Li, J., Zhang, G., Yu, L. et al. [38] provided a cognitive computing system framework aimed at predicting judicial decisions for divorce cases with different expression styles. Sohail et al. [39] implemented ML algorithms on attributes affecting the divorce rate. The authors considered only a few major attributes on a sample dataset, including age, physical/mental abuse, infidelity, and social involvement, and then applied three different classifiers to predict the outcome. Among research papers in the literature, the closest study was presented by Yöntem et al. [40]. The authors investigated the effect of various features on predicting whether the couples' marital status ended with them being either married or divorced. The authors applied artificial neural networks and random forest methods in this two-class classification model. As mentioned earlier, in a country like Iran, where divorce is discouraged, and distressed couples are encouraged to stay married even if they suffer from extremely low marital satisfaction, it can be assumed that many married couples are emotionally separated from each other [41]. This is why the present study puts emphasis on the distinction between happily married and unhappily married (emotionally separated) couples. The purpose of this study (predicting emotional separation and divorce) is pursued by means of artificial intelligence in order to get more accurate results, in the hopes that this can lead to further promising results in regards to applying artificial intelligence to other domains of human behavior.

3 Method

3.1 Measures

The following are data collection instruments that were used in this study:

- Personal information form: This form was developed by the researchers and includes questions on gender, age, education, and marital status.
- Young Schema Questionnaire (YSQ-SF2) [31]: This questionnaire consists of 75 items that examine the existence of 15 early maladaptive schemas. Each schema is assessed by 5 items on a 6-point Likert. Oei and Baranoff [42] achieved the Cronbach's Alpha of YSQ-SF0 as 0.80. The Cronbach's Alpha of YSQ in the present study was measured by SPSS and found to be 0.82.
- Gottman Emotional Separation Scale [43]: Consists of 24 items on a yes/no scale. Any yes equals one point, with the final score being somewhere between 0 to 24. If the final score is 8 or above, the participants are experiencing some level of emotional separation. The Cronbach's Alpha of the Gottman Emotional Separation Scale in the present study was measured by SPSS and found to be 0.84.
- Persian Divorce Prediction Scale (PDPS) four-point Likert scale that examines the couple's relationship when it comes to addressing its most essential points, including love and affection towards each other, shared activities, communication styles, and relationships with families. At first, the researchers conducted twenty-two interviews with twenty-two individuals from different cities of Iran who voluntarily attended the interviews. Among participants, seventeen were married, and five were divorced. The Gottman Emotional separation scale was used to distinguish between "happily married" and "unhappily married" participants. Among participants, seven were found to be emotionally divorced, and eight people were happily married. Fifteen interviews were conducted face-to-face, and seven interviews were carried out online via Skype due to COVID-19 restrictions during the months of study. Each interview lasted 45-55 minutes, and participants were asked to talk about what bothered them in their relationships, answering questions such as "What could change so that you would feel better?". Questions for these interviews were designed by two psychologists based on the literature review, and the question lists were different for each of the three groups. However, the goal of all of the interviews was to answer two questions: what made the happily married couples' relationship satisfying, and what bothered divorced and emotionally separated couples. In the next step, thirty-eight underlying themes were gathered and shared with five couples' therapists in Tehran and four other cities in Iran. The therapists were asked to add any item that they believed should be added to the list, and finally, a 41-item scale was made. A scale draft was then shared with seven married or divorced individuals to ensure that the items were comprehensible. Some items were revised and paraphrased based on the feedback from married and divorced individuals, and finally, the PDPS was ready for use in the study. Cronbach's Alpha coefficient obtained in the present study was 0.87 for this scale

3.2 Analytical Strategy

After collecting survey responses, the data analysis started. Two hundred fifty-nine answers remained after cleaning the data. Of all the participants, 52% were happily married, 25% were emotionally separated, and 23% were divorced. A summary of the participants' demographic information is illustrated in Fig. 1.

Table 1 lists a detailed description of all the features considered in the proposed model. The total input features were 54; it consisted of 15 schema-related questions and

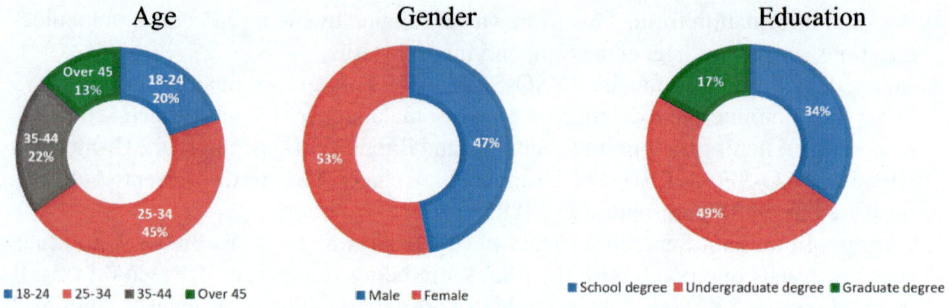

Fig. 1. The percentage of participants by age, gender, and education

39 PDPS questions. There is one output target variable, classifying the couples' marital status into three different class labels (i.e., happily married, emotionally separated, and divorced).

3.3 Methodology

This section presents the steps required to implement machine learning algorithms aimed at achieving the highest possible accuracy in the prediction of marital stability in a real case study. As it can be seen, alongside four classification algorithms (LR, SVM, DT, and RF), two feature selection algorithms Univariate and Relief were employed to select the most important features of the model.

3.4 Data Preprocessing

Data preprocessing is the first crucial step in utilizing classification algorithms efficiently. In this study, after cleaning the data, the percentages of the class labels 1, 2, and 3 were respectively equal to 62.2, 22.0, and 15.8%, which shows imbalanced datasets, as you would expect with real-world cases [44–46]. To address such unbalancing class labels of the target variables, we employed the most widely used approach: a synthetic minority over-sampling technique (SMOTE) [47], which artificially generates new samples from the minority classes using their nearest neighbors. In general, SMOTE provides a more balanced dataset. Furthermore, to achieve a more precise analysis, data transformation was performed by applying a MaxMin scaler, which removes the negative influence of the raw features' value and scales the values in the same range between 0 and 1.

3.5 Feature Selection Algorithms

The chief benefit of utilizing feature selection algorithms is determining the most important attributes. Additionally, this approach can reduce overfitting by eliminating noisy features (i.e., non-informative and redundant data), reducing training time, and boosting prediction accuracy [48]. In this study, two types of feature selection methods, Univariate and Relief, were used alongside considering full features in the prediction of response variables. The univariate feature selection algorithm selects the best attributes based on

Table 1. List of features for prediction of marital stability and divorce

No.	Feature name	No.	Feature name
1	Schema: Emotional deprivation	28	PDPS: Mutual understanding
2	Schema: Abandonment	29	PDPS: Being admired by the spouse for personality
3	Schema: Mistrust/abuse	30	PDPS: Choosing the spouse if made to choose between one's parents and spouse
4	Schema: Social isolation	31	PDPS: Being aroused by the spouse
5	Schema: Defectiveness/shame	32	PDPS: Efforts in the relationship being acknowledged by the spouse
6	Schema: Failure	33	PDPS: Being sure of spouse's fidelity
7	Schema: Dependence/incompetence	34	PDPS: Consensus on how to deal with each other's parents
8	Schema: Vulnerability to harm	35	PDPS: Having time for each other
9	Schema: Enmeshment/underdeveloped self	36	PDPS: Consensus on free time activities
10	Schema: Subjugation	37	PDPS: Families agree with marriage
11	Schema: Self-Sacrifice	38	PDPS: Being there for each other
12	Schema: Emotional inhibition	39	PDPS: Able to compromise easily
13	Schema: Unrelenting standards	40	PDPS: Spouse, the best friend
14	Schema: Entitlement	41	PDPS: Enjoy each other's company
15	Schema: Insufficient self-control	42	PDPS: Respecting each other and avoiding criticism
16	PDPS: The best person in each other's life	43	PDPS: Knowing the spouse will choose me over their parents
17	PDPS: Sexual relationship consensus	44	PDPS: Able to share concerns without fear of judgment or blame
18	PDPS: Awareness of each other's interests	45	PDPS: Sharing the same religious views
19	PDPS: Acceptance from the spouse's family	46	PDPS: Feeling love and compassion for the spouse
20	PDPS: Cooperation on decision making	47	PDPS: Consensus on child-rearing style (if having children)
21	PDPS: Adequate time for getting to know each other before marriage	48	PDPS: Physical aggression
22	PDPS: Aware of each other's worries	49	PDPS: Blaming each other for mistakes
23	PDPS: Husband, as a back for wife	50	PDPS: Harsh criticism and judgment

(*continued*)

Table 1. (*continued*)

No.	Feature name	No.	Feature name
24	PDPS: Knowing how to express love	51	PDPS: Disagreement on financial decisions
25	PDPS: Being considered sexually attractive by the spouse	52	PDPS: Normally being together at the end of arguments
26	PDPS: Knowing the reasons for marrying this person	53	PDPS: Frequency of misunderstandings
27	PDPS: Being proud of the spouse	54	PDPS: Remembering sad memories more than sweet ones

univariate statistical tests. In this research, we utilized the SelectKBest method, which uses a statistical test of chi-squares to select the highest influential k attributes and ignores low-impact and irrelevant attributes for classification. Relief, as another method for feature reduction in the big dataset, is capable of detecting feature dependencies that arise because of assigning weights to all the features in the dataset, which are updated over time [49]. The nearest neighbor concept is used in this algorithm to indirectly represent interactions and select the highest relative features with the largest weight.

3.6 Supervised Machine Learning Algorithms

Machine learning is a subset of artificial intelligence that studies computer algorithms aimed at mining complex datasets, parsing data, learning a task from experience, and then deciding or predicting with a high degree of accuracy. Supervise ML (SML) algorithms are a type of ML that creates predictive models from already existing classified samples and labeled data [50]. In the current study, four SML classifiers, namely SVM, LR, DT, and RF, were used to test and ultimately find the optimal classifier that achieves the highest accuracy for divorce prediction. Below, these employed SML classifiers are briefly presented.

- SVM is a predictive modeling approach that has been used for both regression and classification problems [51]. The SVM algorithm aims to construct a set of hyperplanes, which divides the data into several classification groups by boundaries. Each hyperplane has the greatest margin to the edge of each group. A subset of the training set (i.e., supported vectors) is constructed to define the boundaries. Then, the SVM method seeks the optimal separating hyperplane with the best generalization performance attained via maximizing the distance between the margins and hyperplane. To take data as input and transform it into a mandatory form, SVM uses different kernel functions such as linear, polynomial, sigmoid, and radial [52].
- LR is another supervised machine learning algorithm that looks to find a monotonous differentiable function (i.e., Sigmoid). Not only can LR classify a test sample, but it also provides the corresponding probability that is the most beneficial for supporting a decision. Generally, response variable y in logistic regression is dichotomous,

which means there would be only two possible classes [0,1]. However, in ordinal and multimodal types of logistic regression, target variables can have three or more target variables.

- DT is the third classification algorithm utilized in this study. This training algorithm has a tree structure consisting of a collection of nodes, leaf nodes, and branches, representing attributes, class labels, and outputs of the test, respectively [53]. In this algorithm, the data is split into two or more homogeneous sets based on a measure that maximizes the separation of the data. The main advantages of this method compared to other algorithms are: 1) DT does not require normalization and scaling of data, which results in shortening preprocessing time; and 2) it uses a relatively simple format, helping decision-makers to interpret it straightforwardly.
- Finally, RF is an ensemble supervised learning classification algorithm composed of a combination of three classifiers developed by Breiman [54]. The difference between RF and DT is that a DT is built on an entire dataset using all attributes; however, RF randomly selects a bunch of samples from the training data, chooses a random set of attributes to construct multiple DTs, and finally aggregates the set of estimated DTs to get a single decision tree. Therefore, the central idea of RF is to create several DTs on random subsets.

3.7 Tuning the Hyperparameter

The performance of the classification models is quite sensitive to the value of the hyperparameters of the training algorithms. Thus, the hyperparameters should be tuned to achieve the best combination that yields high performance. We considered different levels for each hyperparameter of the learning algorithms. The Taguchi method [55] and grid search technique can be best utilized to tune these parameters. The former is applied to reduce the number of experiments and determine an acceptable value for each parameter by maximizing the signal-to-noise ratio. However, the latter considers all permutations regarding the full factorial design and is always necessary to determine the optimal value for each hyperparameter [52]. In this study, we used the grid search approach to maximize the overall accuracy of the classification models since it can obtain the optimal value of hyperparameters in a reasonable computational time.

Furthermore, to obtain reliable estimates and avoid an overfitting issue, a k-fold cross-validation technique was performed. To do so, the training dataset was partitioned into k independent folds of equal size, in which $k-1$ groups are used for model training, and one-fold is used for validating the trained models. This process is repeated k times, and the average accuracy of the models is calculated by looking at which achieves the best accuracy for the models. In our experiment, we considered three different levels for CV; 5, 10, and 15. Table 2 lists the hyperparameters and their levels in ML algorithms. As is shown, the tuning hyperparameters of each algorithm are as follows: 1) SVM: regularization (Reg), Kernel, Gamma (G); 2) LR: penalty, solver, and reg; 3) DT: The minimum required samples to split (MS), the minimum required samples to be at a leaf node (ML), criterion (Cri), and Splitter; 4) RF: Number of trees (NT), MS, ML, and Cri. The best combination of parameters for each classifier is reported in Table 3.

Table 2. The value of the hyperparameter corresponds to each level for different ML algorithms.

Level	SVM			LR			DT				RF			
	Reg	Gamma	Kernel	Reg	Solver	Penalty	MS	ML	Splitter	Cri	MS	ML	NT	Cri
1	0.001	0.01	Linear	1	NC*	L1	2	1	Best	Ent*	2	1	50	Ent
2	0.010	0.10	Poly*	5	LBFGS*	L2	3	2	Random	Gini	3	2	100	Gini
3	0.100	1.00	RBF*	10	SAG	–	4	3	–	–	4	3	150	–
4	1.000	10.0	Sigmoid	50	SAGA	–	5	4	–	–	5	4	200	–
5	10.00	100	–	100	–	–	6	5	–	–	6	5	300	–

Poly: Polynomial; RBF: Radial Basis Function; NC: Newton-CG; LBFGS: Limited memory BFGS; Ent: Entropy.

Table 3. The best value of the hyperparameter corresponding to each level for different ML algorithms.

SVM			LR			DT				RF			
Reg	Gamma	Kernel	Reg	Solver	Penalty	MS	ML	Splitter	Cri	MS	ML	NT	Cri
10.0	0.01	RBF	5	SAG	L2	2	1	Random	Gini	2	1	100	Gini

3.8 Evaluation Metrics

The performances of employed multi-class classification algorithms were assessed using the typical performance measures for data classification [56]. In this study, we applied the accuracy metrics to tune the hyperparameters and compare the performance of the predictors. Additionally, performance measures including Precision, Recall (also called sensitivity), and F1-score for each class were used to gain more insight into the results and thoroughly evaluate the classifiers.

4 Results

4.1 Experimental Setup

The classification algorithms were implemented in Python 3.8. All algorithms were executed on a computer with a CPU Core i7-6700 and 16 GB memory using the Windows 10 operating system. Moreover, we used various Python libraries both for feature selection and labeling the classification task concerning the prediction of divorce.

In this study, alongside considering full features in the model, we also utilized two feature selection methods: Univariate and Relief. The goal was to select the most important attributes and evaluate the performance results of these three approaches to obtain the maximum accuracy in predicting marital stability. Ten attributes were included in both algorithms among the fifteen most important features. It can be noted that in both algorithms, "my spouse is my best friend" and "there is a mutual understanding between us" are the most influential attributes in such a way that the former has the highest score in the Univariate method, and the latter has the highest rank in Relief. Three schemas (emotional deprivation, mistrust/abuse, and enmeshment) are among the top fifteen features in terms of accurately predicting divorce. Emotional deprivation is concerned with feeling understood, supported, and taken care of. By looking at some other items of PDPS that are among the top fifteen features (e.g., "no matter how busy we are, we can always have some time for each other," and "we know how to express our love to each other"), we can see that these items are all related to this very schema. Suppose a wife or husband feels their spouse is always there for them and knows they can express and receive love from them easily. In that case, a secured bond can evolve between them, which heals the possible worries regarding the emotional deprivation schema. The mistrust/abuse schema refers to a profound and chronic feeling of insecurity and a lack of trust in others.

Since this schema is also concerned with interpersonal feelings of trust and security, it is no surprise that, among other schemas, this is one of the three most significant predictors of marital stability. Enmeshment is the third schema in the list of the top fifteen features. This schema refers to the inability of the individual to make decisions, act, feel, and understand distinguished from their parents. Enmeshment can prevent the spouse from what Bowen calls self-differentiation [57]. Self-differentiation from the family of origin is thought to be a significant factor in psychological functioning [57]. This issue is even more important when we notice that one of the significant problems of many couples in Iran is how they handle the interferences of families of origin [58]. The less differentiated the spouse, the more families can interfere with the couple's life and decisions. It is then understandable that the enmeshment schema challenges the differentiation of the spouse and makes them vulnerable to this risk factor, at least regarding these Iranian couples' relationships. A husband or wife suffering from an undeveloped self may be unable to support their spouse in a time of distress and/or conflict with their families of origin and may lack adequate psychological functioning in the relationship. The results of feature selection methods that were used in machine learning algorithms are presented in the next section.

4.2 Results of Utilizing Classification Algorithms

This section focuses on evaluating the performance of the four machine learning algorithms employed in this study (i.e., SVM, LR, DT, and RF). The model was implemented by considering both a set of full features alongside the features selected by the Univariate and Relief methods. To do so, we used the cross-validation and hyperparameter tuning results presented in Table 3. Keeping in line with common practices in the literature, we partitioned the dataset into 75% and 25% for the training and testing sets, respectively. Thus, after compensating for unbalanced classes by adopting SMOTE to generate new samples, our dataset included 270 training datasets and 90 test datasets.

Figure 2 illustrates the overall accuracy of the learning algorithms applied to the set of full features alongside the selected features by Univariate and Relief methods. As can be seen, the average accuracy of the predictive algorithms after employing the feature selection techniques slightly improved. The average ACC of classifiers (i.e., mean of ACCs) for the set of full features was 86.1%, and with Univariate and Relief, the accuracy increased to 86.5% and 87.0%, respectively. From the classifiers' perspective, the predictive models of SVM and RF achieved the highest average ACC using three different sets of features at 88.2%, while using such measures for LR and DT gave us 87.1% and 82.6%, respectively. However, the maximum accuracy was observed at 89.2% for SVM and RF. Since the frequency of finding the best accuracy was found in the model under a set of full features, in the following, we assess other performance measures of classifiers by taking the full features into account. To this end, we used the confusion matrix (also called error matrix) as a visual evaluation tool shown in Table 4. The actual (also called truth) class results are represented in the rows of the confusion matrix, and the prediction class results are in the columns. The diagonal elements of the confusion matrix represent the true positive values in each class.

As can be observed from the confusion matrix in Table 4, SVM and RF learning algorithms performed better in classes 1 and 2 in regards to Precision, Recall, and F1-Score measures. Regarding class 3, the performance of LR overshadowed all other algorithms. Despite achieving a high level of accuracy at 80%, the DT predictive model

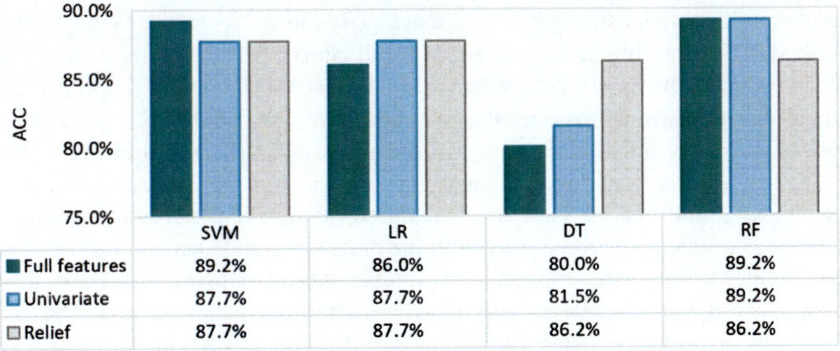

Fig. 2. The accuracy of the learning algorithms applied on a set of full features and selected features by univariate and relief methods

showed the lowest performance in the accurate prediction of all classes. Altogether, SVM and RF predictive learning models reported the highest performance among the classifiers. Although the overall accuracy of both SVM and RF was about 89%, RF showed slightly better performance, thanks to its higher accuracy in the prediction of class 3; it has shown itself to be the superior algorithm for this study.

Table 4. The confusion matrix and performance measures of different types of classifiers involving a set of full features

Prediction →	SVM			LR			DT			RF		
Actual ↓	1	2	3	1	2	3	1	2	3	1	2	3
1	41	0	0	36	5	0	38	2	1	41	0	0
2	0	12	2	0	13	1	1	10	3	1	12	1
3	0	5	5	0	3	7	1	5	4	0	5	5
Precision	1.00	0.71	0.71	1.00	0.62	0.88	0.95	0.59	0.50	0.98	0.71	0.83
Recall	1.00	0.86	0.50	0.88	0.93	0.70	0.93	0.71	0.40	1.00	0.86	0.50
F1-score	1.00	0.77	0.59	0.94	0.74	0.78	0.94	0.65	0.44	0.99	0.77	0.62
Accuracy	**0.892**			**0.862**			**0.800**			**0.892**		

Accuracy—another performance measure that we focused more on to compare the machine learning algorithms. The Accuracy represents the number of correctly classified data instances over the total number of data instances

5 Discussion

Increasing divorce rates have become a major concern worldwide, especially in Iran. The present study aimed to develop a tool for predicting marital stability among Iranian couples. This study presented an intelligent decision support system to predict whether a couple would separate using supervised machine learning algorithms. With that aim, a scale called PDPS was first developed after several interviews with married and divorced people and couples therapists. Afterward, the marital status of couples (happily married, emotionally separated, or divorced) was studied based on YSQ-SF2 and PDPS. Two methods, Univariate and Relief, were utilized to extract the most important features in the dataset. Several machine learning algorithms, including SVM, LR, DT, and RF, were equipped and tested on a set of full-featured (alongside the selected features) for their ability to predict marital stability with the highest possible accuracy. Furthermore, grid search and k-fold cross-validation techniques were employed to achieve the best combination of hyperparameters that yielded a high performance. Overall, the highest accuracy was achieved by using full features and SVM and RF at 89.2%. However, when considering various metrics as they pertained to predicting different class labels, RF performed slightly better overall. Additional findings implied that couples who can express their emotions, feel like friends with each other, are available to each other, and can handle the possible interferences of families of origins for themselves and their spouse, were less likely to experience emotional or formal divorce. By applying the findings of this study, couples who are in danger of separation can be alerted to this danger earlier on and make the necessary changes, if possible, to separation divorce and its detrimental consequences.

Although this study addressed some gaps in the literature, some limitations still need to be tackled in the future. From the modeling side, when it comes to evaluating the early maladaptive schemas among participants, conducting interviews alongside questionnaires could provide more accurate information. From a methodology perspective, other feature selection techniques and supervised machine learning algorithms should be utilized, and then the performance metrics should be compared with the presented work. Additionally, to boost artificial intelligence systems' capabilities, the outputs of multiple classifiers could be combined and integrated into an ultimate decision via a voting process [13]. While considering this strategy, the prediction accuracy for both the selected features and the full features' set can be further investigated to obtain both the best performance and certainty. Finally, to better interpret the results, it is recommended that instead of relying on purely statistical machine learning performance, the explainability results be examined.

References

1. Singer, L.J.: Divorce and the single life: divorce as development. J. Sex Marital Ther. **1**(3), 254–262, n, 1–8 (1975)
2. Robles, T.F., Slatcher, R.B., Trombello, J.M., McGinn, M.M.: Marital quality and health: a meta-analytic review. Psychol. Bull. **140**(1), 140 (2014)
3. Segraves, R.T.: Marriage and mental health. J. Sex Marital Ther. **6**(3), 187–198 (1980)
4. Tumin, D.: Does marriage protect health? A birth cohort comparison. Soc. Sci. Q. **99**(2), 626–643 (2018)
5. Kapelle, N., Baxter, J.: Marital dissolution and personal wealth: examining gendered trends across the dissolution process. J. Marriage Fam. **83**(1), 243–259 (2021)
6. Goisis, A., Özcan, B., Van Kerm, P.: Do children carry the weight of divorce? Demography **56**(3), 785–811 (2019)

7. Haimi, M., Lerner, A.: The impact of parental separation and divorce on the health status of children, and the ways to improve it. J. Clin. Med. Genomics **4**(1), 137 (2016)
8. Amato, P.R., Cheadle, J.: The long reach of divorce: Divorce and child well-being across three generations. J. Marriage Fam. **67**(1), 191–206 (2005)
9. Iranian Civil Registration. Statistics. Retrieved from https://www.sabteahval.ir (2020)
10. Greenwood, J., Guner, N., Kocharkov, G., Santos, C.: Technology and the changing family: A unified model of marriage, divorce, educational attainment, and married female labor-force participation. Am. Econ. J. Macroecon. **8**(1), 1–41 (2016)
11. Cito, G., et al.: The impact of the COVID-19 quarantine on sexual life in Italy. Urology **147**, 37–42 (2021)
12. Hansen, L.B., Shireman, J.F.: The process of emotional divorce: examination of theory. Soc. Casework **67**(6), 323–331 (1986)
13. Hansen, L.K., Salamon, P.: Neural network ensembles. IEEE Trans. Pattern Anal. Mach. Intell. **12**(10), 993–1001 (1990)
14. Sahebihagh, M.H., Khorshidi, Z., Atri, S.B., Jafarabadi, M.A.: Investigating the relationship between self-efficacy and emotional divorce among nurses in the city of Rasht, Iran (2015). Ann. Trop. Med. Public Health **10**(6), 1596 (2017)
15. Hashemi, L., Homayuni, H.: Emotional divorce: child's well-being. J. Divorce Remarriage **58**(8), 631–644 (2017)
16. Karlsen, A.S., Futris, T.G., Richardson, E.W.: The dyadic effects of relationship uncertainty on relationship maintenance and damaging behaviors. J. Couple Relat. Ther. Advance online publication (2020). https://doi.org/10.1080/15332691.2020.1837323
17. Shahraki, G., Sedaghat, Z., Fararouei, M., Tabeshfar, Z.: Is Divorce Predictable among Iranian Couples? Can. J. Fam. Youth [Le Journal Canadien de Famille et de La Jeunesse] **12**(1), 243–259 (2020)
18. Amato, P.R., Beattie, B.: Does the unemployment rate affect the divorce rate? An analysis of state data 1960–2005. Soc. Sci. Res. **40**(3), 705–715 (2011)
19. Bernardi, F., Martínez-Pastor, J.-I.: Divorce risk factors and their variations over time in Spain. Demogr. Res. **24**, 771–800 (2011)
20. Jensen, P., Smith, N.: Unemployment and marital dissolution. J. Popul. Econ. **3**(3), 215–229 (1990)
21. Hohmann-Marriott, B.E., Amato, P.: Relationship quality in interethnic marriages and cohabitations. Soc. Forces **87**(2), 825–855 (2008)
22. Song, M.: Does 'race' matter? A study of 'mixed race' siblings' identifications. Soc. Rev. **58**(2), 265–285 (2010)
23. Fardis, M.: Expression and regulation of emotions in romantic relationships (2007)
24. Sbarra, D.A., Hasselmo, K., Bourassa, K.J.: Divorce and health: beyond individual differences. Curr. Dir. Psychol. Sci. **24**(2), 109–113 (2015)
25. Solomon, B.C., Jackson, J.J.: Why do personality traits predict divorce? Multiple pathways through satisfaction. J. Pers. Soc. Psychol. **106**(6), 978 (2014)
26. Gottman, J.M.: What Predicts Divorce?: The Relationship Between Marital Processes And Marital Outcomes. Psychology Press (2014)
27. Story, L.B., Karney, B.R., Lawrence, E., Bradbury, T.N.: Interpersonal mediators in the intergenerational transmission of marital dysfunction. J. Fam. Psychol. **18**(3), 519–529 (2004). https://doi.org/10.1037/0893-3200.18.3.519. PMID: 15382977
28. Afshani, S.A., Ardian, N., Sorbi, M.H., Mazloomy-Mahmoodabad, S.S., Morowati-Sharifabad, M.A., Goodarzi-Khoigani, M.: A theory-based educational intervention impact on the reduction of divorce intention in Iran. Curr. Psychol. 1–8 (2020)
29. Yoosefi, N., Etemadi, O., Bahrami, F., Fatehizade, M.A.S., Ahmadi, S.A.: An investigation on early maladaptive schema in marital relationship as predictors of divorce. J. Divorce. Remarriage. **51**(5), 269–292 (2010)

30. Mohammadi, B., Soleymani, A.: Early maladaptive schemas and marital satisfaction as predictors of marital commitment. Int. J. Behav. Sci. **11**(1), 16–22 (2017)
31. Young, J.E., Klosko, J.S., Weishaar, M.E.: Schema therapy. New York: Guilford, 254 (2003)
32. Waheed, H., Hassan, S.-U., Aljohani, N.R., Hardman, J., Alelyani, S., Nawaz, R.: Predicting academic performance of students from VLE big data using deep learning models. Comput. Hum. Behav. **104**, 106189 (2020)
33. Xu, X., Wang, J., Peng, H., Wu, R.: Prediction of academic performance associated with internet usage behaviors using machine learning algorithms. Comput. Hum. Behav. **98**, 166–173 (2019)
34. Posada Trobo, I., García Díaz, V., Pascual Espada, J., González Crespo, R., Moreno-Ger, P.: Rapid modeling of human-defined AI behavior patterns in games. J. Ambient Intell. Humaniz. Comput. **10**(7), 2683–2692 (2018). https://doi.org/10.1007/s12652-018-0969-y
35. Igwe, O.M., Wang, Y., Giakos, G.C., Fu, J.: Human activity recognition in smart environments employing margin setting algorithm. J. Ambient Intell. Humaniz. Comput. 1–13 (2020)
36. Chatterjee, A., Gupta, U., Chinnakotla, M.K., Srikanth, R., Galley, M., Agrawal, P.: Understanding emotions in text using deep learning and big data. Comput. Hum. Behav. **93**, 309–317 (2019)
37. Guillén, A., Tovar, C., Herrera, L.J., Pomares, H., González, J., Guillén, J.F., Rojas, I.: Predicting marital dissolutions using radial basis function neural networks. In: The 2010 International Joint Conference on Neural Networks (IJCNN), pp. 1–4 (2010)
38. Li, J., Zhang, G., Yu, L., Meng, T.: Research and design on cognitive computing framework for predicting judicial decisions. J. Signal Process. Syst. **91**(10), 1159–1167 (2018). https://doi.org/10.1007/s11265-018-1429-9
39. Sohail, S., Aziz, S., Tahir, F., Haqqui, S., Hussain, A.: Implementation of machine learning algorithm on factors effecting divorce rate. In: 2018 International Conference on Engineering and Emerging Technologies (ICEET), pp. 1–5 (2018)
40. Yöntem, M.K., Kemal, A.D.E.M., Ilhan, T., Kiliçarslan, S.: Divorce prediction using correlation based feature selection and artificial neural networks. Nevşehir Hacı Bektaş Veli Üniversitesi SBE Dergisi. **9**(1), 259–273 (2019)
41. Barzoki, M.H., Tavakoll, M., Burrage, H.: Rational-emotional 'divorce'in Iran. Appl. Res. Qual. Life **10**(1), 107–122 (2015)
42. Oei, T.P., Baranoff, J.: Young schema questionnaire: review of psychometric and measurement issues. Aust. J. Psychol. **59**(2), 78–86 (2007)
43. Gottman, J., Gottman, J.M., Silver, N.: Why Marriages Succeed or Fail: and How You Can Make Yours Last. Simon and Schuster (1995)
44. Aghaaminiha, M., Ghanadian, S.A., Ahmadi, E., Farnoud, A.M.: A machine learning approach to estimation of phase diagrams for three-component lipid mixtures. Biochim. Biophys. Acta (BBA)-Biomembr. 183350 (2020)
45. Cai, Q., Abdel-Aty, M., Yuan, J., Lee, J., Wu, Y.: Real-time crash prediction on expressways using deep generative models. Trans. Res. Part C: Emerg. Technol. **117**, 102697 (2020)
46. Chen, K.-H., Wang, K.-J., Adrian, A.M., Wang, K.-M., Teng, N.-C.: Diagnosis of brain metastases from lung cancer using a modified electromagnetism like mechanism algorithm. J. Med. Syst. **40**(1), 35 (2016)
47. Chawla, N.V., Bowyer, K.W., Hall, L.O., Kegelmeyer, W.P.: SMOTE: synthetic minority over-sampling technique. J. Artif. Intell. Res. **16**, 321–357 (2002)
48. Adeli, E., Li, X., Kwon, D., Zhang, Y., Pohl, K.: Logistic regression confined by cardinality-constrained sample and feature selection. IEEE Trans. Pattern Anal. Mach. Intell. (2019)
49. Urbanowicz, R.J., Meeker, M., La Cava, W., Olson, R.S., Moore, J.H.: Relief-based feature selection: introduction and review. J. Biomed. Inform. **85**, 189–203 (2018)

50. Hwong, Y.-L., Oliver, C., Van Kranendonk, M., Sammut, C., Seroussi, Y.: What makes you tick? The psychology of social media engagement in space science communication. Comput. Hum. Behav. **68**, 480–492 (2017)
51. Nayak, J., Naik, B., Behera, H.: A comprehensive survey on support vector machine in data mining tasks: applications and challenges. Int. J. Database Theory Appl. **8**(1), 169–186 (2015)
52. Bhadra, T., Bandyopadhyay, S., Maulik, U.: Differential evolution based optimization of SVM parameters for meta classifier design. Procedia Technol. **4**, 50–57 (2012)
53. Rokach, L., Maimon, O.: Top-down induction of decision trees classifiers-a survey. IEEE Trans. Syst. Man, Cybern. Part C (Appl. Rev.) **35**(4), 476–487 (2005)
54. Breiman, L.: Random forests. Mach. Learn. **45**(1), 5–32 (2001)
55. Taguchi, G.: System of experimental design; engineering methods to optimize quality and minimize costs (1987)
56. Sokolova, M., Lapalme, G.: A systematic analysis of performance measures for classification tasks. Inf. Process. Manage. **45**(4), 427–437 (2009)
57. Murdock, N.L., Gore, P.A.: Stress, coping, and differentiation of self: a test of Bowen theory. Contemp. Fam. Ther. **26**(3), 319–335 (2004)
58. Barikani, A., Ebrahim, S.M., Navid, M.: The cause of divorce among men and women referred to marriage and legal office in Qazvin, Iran. Global J. Health Sci. **4**(5), 184 (2012)

Predicting the Loading Parameters of a Square Panel Upon Linear Deflection

Leonardo Espinosa-Leal[1(✉)], Silas Gebrehiwot[2,3], and Heikki Remes[3]

[1] Department of Business Management and Analytics, Arcada University of Applied Sciences, Jan-Magnus Janssons plats 1, 00560 Helsinki, Finland
leonardo.espinosaleal@arcada.fi

[2] Department of Energy and Materials Technology, Arcada University of Applied Sciences, Jan-Magnus Janssons plats 1, 00560 Helsinki, Finland
silas.gebrehiwot@arcada.fi

[3] Aalto University, School of Engineering, Department of Mechanical Engineering, P.O. Box 15300, FIN-00076 Aalto, Finland
heikki.remes@aalto.fi

Abstract. This paper presents a machine learning methodology for predicting the static deflection of an isotropic squared panel in the linear-elastic regime. Our method uses a series of simulations via FEM (Finite Element Method) for obtaining the contour plot images of flexural deflections due to point loads at different locations on the panel. From here, we obtain, using sliding windows of different sizes, segments of images that are transformed into features using a state-of-the-art pre-trained deep neural network. We trained three different machine learning models for multi-output regression. Our results show that it is possible to predict with a high score the coordinates of the point of load and the amount of deflection with partial visual information.

Keywords: Mechanical deflection · Deep learning · Extreme Learning Machine

1 Introduction

Predicting the mechanical properties of materials *in situ* is a challenging task. In particular, if the material is already in use. In this context, Digital Image Correlation (DIC) has surged as a powerful technique to predict the mechanical properties (e.g., stress and strain). The obtained information can be translated quickly into Deflection Contour Plots (DCP), providing reliable and valuable mechanical information about structural displacements under load. In this context, one important question is the localization of the region of load, where the concentration of stress can give rise to failure in the material. Therefore, finding the exact location of this zone can be of paramount importance, for instance, fixing possible damages or reinforcing the possible areas of failure.

K.-M. Björk (Ed.): ELM 2021, PALO 16, pp. 84–91, 2023.
https://doi.org/10.1007/978-3-031-21678-7_8

Machine learning methods have become a ubiquitous technology in all areas of knowledge. Their use in the field of materials science at different scales has created a new field of knowledge named *materials informatics* [1,13]. In this work, we study the deflection on the linear-elastic regime of one isotropic squared panel using a combination of simulations, first by solving the deformation equations using the Finite Element Method (FEM). We then use the obtained information within a deep learning network for feature extraction, furthermore different machine learning models to predict the coordinates of the point of load and the value of maximum deflection.

1.1 Related Works

The mathematical modeling of the deflection of rectangular panels has been well known since the seminal works by *March* [8] and *Timoshenko* [17,18]. The analytical form is well known, but the solution needs to be approximated using different numerical strategies. In a modern context, computer vision algorithms have been extensively used in the study of mechanical properties of materials, and more importantly in structural engineering applications, for example, civil infrastructure inspection and monitoring [16], stress field prediction [11], prediction of stress-strain curves [19], concrete structural integrity [14], among others [7]. For a more substantial revision of the topic of deep learning in construction and future challenges and applications, see reviews in [9,15] and references therein.

1.2 Deflection of Thin Plates

A plate is defined as thin, when its thickness, h is considerably smaller than its width w and length l. Figure 1 shows the schematic of a thin plate.

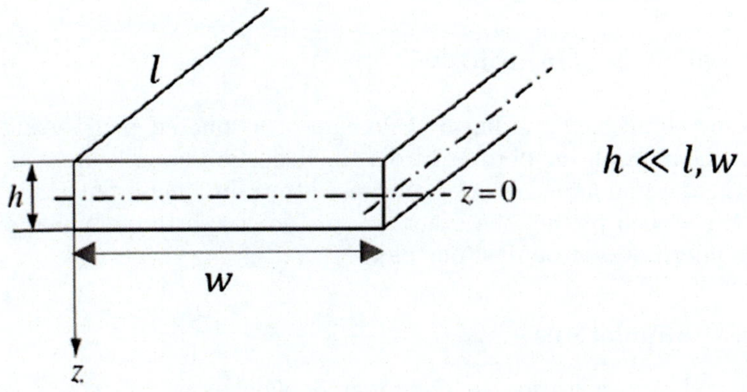

Fig. 1. Schematic of a thin plate

A variety of theories has been developed to govern the deflection of thin walled plate. They vary on the assumptions considered on the orientations of planes regarding the plate's central plane before and after deformations. Under a perpendicular loading, the curvature radii of the plate, ρ_x, ρ_y can be presented in relation to its vertical deformation (see Eq. 1) [10].

$$\frac{1}{\rho_x} = \frac{\partial^2 w}{\partial x^2}, \quad \frac{1}{\rho_y} = \frac{\partial^2 w}{\partial y^2} \tag{1}$$

A thin plate is said to have a cylindrical deflection when only two of its parallel edges are constrained. In a cylindrical deflection where the edges of a thin plate are constrained in y-axis, $\varepsilon \neq 0$. From the plate's curvature radius, the strain can be derived as

$$\varepsilon_x = z\frac{d^2 w}{dx^2} \tag{2}$$

Since, $\varepsilon_y = 0$ and considering the $\sigma_z = 0$ in the $x - z$ plane, the linear elastic constitutive reduces to

$$\varepsilon_x = \frac{1 - v^2}{E}\sigma_x \tag{3}$$

Combining Eqs. 2 and 3, the moment in x can be determined by assuming a unit length

$$M_x = \frac{-Eh^3}{12(1 - v^2)}\frac{d^2 w}{dx^2} \tag{4}$$

In Eq. 4, the $\frac{-Eh^3}{12(1-v^2)}$ term is the bending stiffness of the plate. The deflection of the thin plate in z direction can be determined by solving the above second order differential equation using the boundary conditions. The bending moment depends on the external surface or body forces considered.

2 Research Methodology

We start by simulating the linear deflection of a squared panel using a FEM strategy. The information obtained from the deflection upon loading on a point, in the form of visual information, is then used to train different machine learning models. We aim to predict the exact point of load and the maximal deflection using the information from contour plots.

2.1 FEM Simulations

The linear elastic deflection of the squared panel is calculated by using the mechanical module available on COMSOL multiphysics software [5]. We used a physics controlled meshing with a free tetrahedral meshing and extremely fine element size with a minimum element size of 0.002 mm. The panel has a fixed constraint at its two edge surfaces. We simulate the loading on ten different points on the panel. The boundary a point load which is given as prescribed

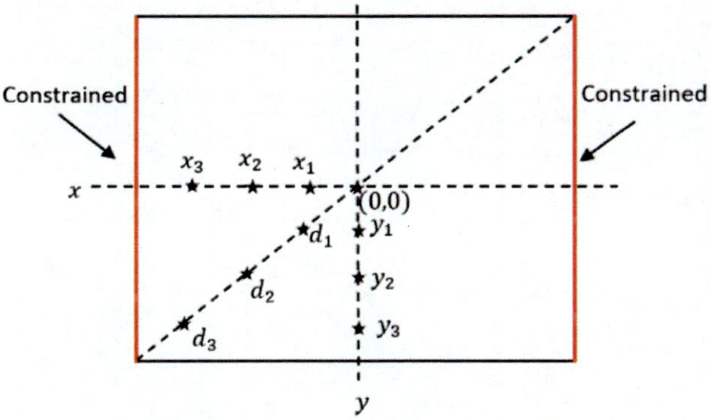

Coordinate	x (mm)	y (mm)
c =(0,0)	0	0
x_1	12.5	0
x_2	25	0
x_3	37.5	0
y_1	0	12.5
y_2	0	25
y_3	0	37.5
d_1	12.5	12.5
d_2	25	25
d_3	37.5	37.5

Fig. 2. *Left:* Diagram that shows the constrained points on the panel (in red) and the position of the ten points of load explored during the simulations. *Right:* Table with the values.

displacement and is linearly related to time. Hence, we know at each time the increment made, how the panel deflection is distributed to all sections of its geometry. Here a 1 mm deflection at each point is made during 100 sec. with load increments made in prescribed displacement of 0.01 mm at each second. That is, displacement = 0.01 mm × time. The loading points and the selected coordinates are shown in Fig. 2. We recorded the images at each time step with a size of 1000 × 1000 pixels. In total, our dataset consists of 1000 images. In Fig. 3 is plotted the deformation contour plots obtained using FEM, at four different points with four different deflection values.

2.2 Machine Learning Modeling

For increasing the predictive power in the trained models, we augmented the information from the dataset by using sliding windows of different sizes with no overlapping. The size of these windows was changed in powers of 2^n where $n = 6, 7, 8, 9, 10$ pixels of side. For avoiding issues with the sizes, the original images were processed to obtain a final size of 1024 pixels per side instead of 1000 pixels. The use of sliding windows for augmenting visual data has been shown highly effective and efficient in other computer vision applications [3,4]. Furthermore, we used the pre-trained deep neural network *Inception21k* [6] to obtain a vector of features from each element in the subset of sliced images (see Fig. 4 for a simple scheme of the research pipeline).

The final data is represented as matrix of dimension $1024 \times 1000 * \left[\frac{1024}{2^n}\right]^2$, which means, for $n = 6$, the windows were of 64×64 pixels^2 and $n = 10$ means the full image (or window of 1024×1024 pixels^2). For prediction, we trained three different machine learning models for multioutput regression, two by using the standard python library *scikit-learn* [12]: Ridge and Random Forest (RF), and the third a new compatible implementation of Extreme Learning

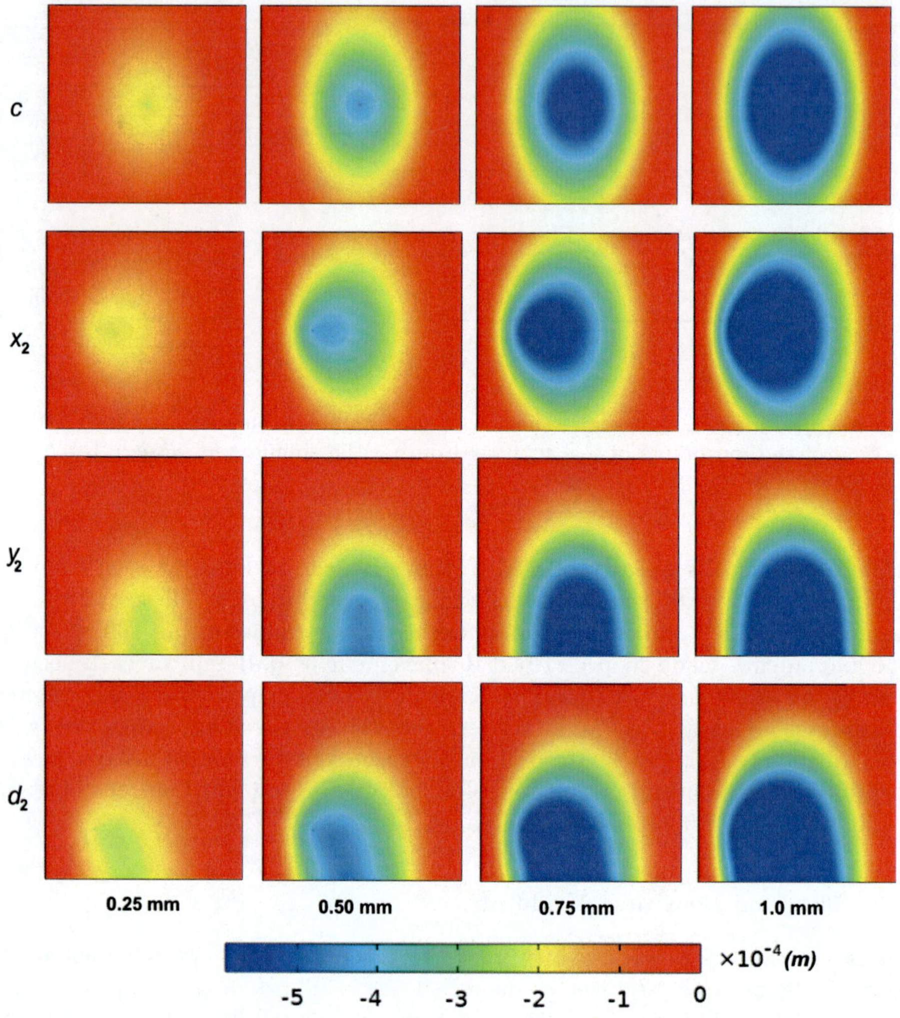

Fig. 3. Contour plots showing obtained by FEM simulations using COMSOL, in the linear deformation regime, upon load at four different points: $c = (0, 0)$, x_2, y_2 and d_2, at four different maximum deflection values (0.25 mm, 0.50 mm, 0.75 mm and 1.0 mm).

Machine (ELM) named *scikit-elm* [2]. The target is a three-dimensional vector with the value of the linear deflection at the point of load and its 2D coordinates (see Fig. 4). For obtaining the best model, we performed a random grid search of parameters of 1000 steps with R^2 as a scoring metric and five-fold cross-validation. The obtained results for R^2 with the respective optimized parameters are summarized in Table 1. In the case of the Ridge model, we varied the L2 regularization parameter α in the interval $[10^{-5}, 10^3]$, 1000 values. For the ELM

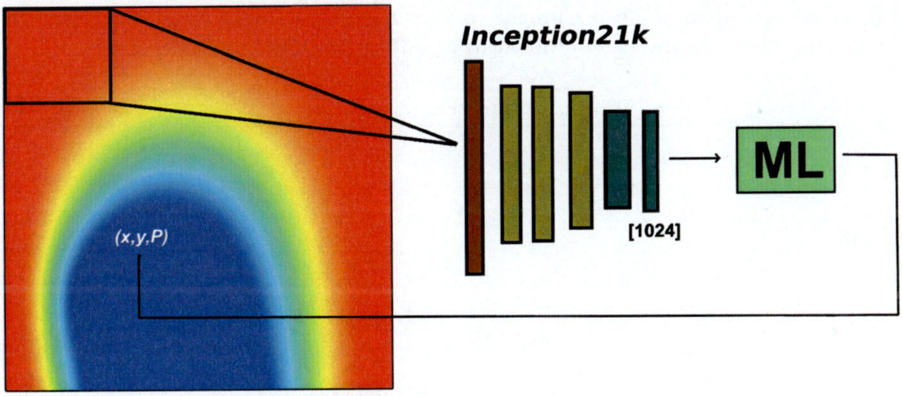

Fig. 4. Schematic representation of the machine learning pipeline.

calculations we optimized the number of neurons (N), the L2 regularization parameter $\alpha = [10^{-5}, 10^{3}]$, the density of input weights (similar to dropout rate in neural networks) in the interval $[0.01, 0.99]$ and the mathematical form of the transformation function $f = \{\text{ReLU, Linear, Sigmoid, Hyperbolic tangent}\}$.

3 Results and Discussion

The coordinates on the plane and the deepness of the deflection in the linear regime can be predicted with high accuracy using partial visual information. In Table 1, a comparison among the three machine learning models is presented. The three models give a similar result when the features are extracted from the full-size images (1024 pixels $\times$ 1024 pixels). In the case of ELM, the random search with the best score is when the neurons are linear. When the size of the sliding windows decreases, the accuracy of Ridge decreases in comparison with ELM. Because of time limitations, the calculations done using random forest were only limited to full-size images. All three methods produced the same score for the full-size images, 0.97. Moreover, because the images were obtained in the linear regime, the problem for this data is linear. For instance, the neurons obtained after optimizing the parameters were linear (lin) for the ELM case.

From the obtained results, the score for the ELM results is slightly better than Ridge at larger sizes of sliding windows, which means the models are learning the linearity of the data. Moreover, ELM results are superior to Ridge at small sliding windows, and the ReLU function is the optimal transformation function. Besides, the density of input weights is reduced with the inverse of the size of the sliding window. A similar behavior follows the regularization parameter.

4 Conclusion and Future Research

This work showed that it is possible to predict efficiently the linear deflection of one object using visual information. Our results also show that partial infor-

Table 1. Results of the R^2 score on the test set from three machine learning models, including the main parameters obtained from a 1000 steps random grid-search.

n	Ridge	Random forest regression	ELM
6	0.56 ($\alpha = 1e-05$)	–	0.69 N = 3998, $\alpha = 0.007$, f = ReLU, density = 0.01
7	0.67 ($\alpha = 1e-05$)	–	0.78 N = 3884, $\alpha = 1.11$ f = ReLU, density = 0.026
8	0.85 ($\alpha = 1e-05$)	–	0.91 N = 3189, $\alpha = 9.21$, f = ReLU, density = 0.04
9	0.95 ($\alpha = 10.10$)		0.97 N = 2441, $\alpha = 5.05$ f = ReLU, density = 0.02
10	0.97 ($\alpha = 10.10$)	0.97 estimators = 600, max-depth = 60, min-samples-split= 2, max-features= sqrt	0.97 N = 2899, $\alpha = 0.68$, f = linear, density = 0.94

mation, to a certain degree, is enough for predicting purposes. In our case, the contour plots are obtained after post-processing of the numerical data matrices, which means instead of the visual plots, we could use these for training the models. However, visualizing deformation and using these as input for a machine learning regressor is more valuable because it can be easily adapted to current digital image correlation devices.

A comparison among the different models shows that ELM performs superiorly in terms of score and computational performance, including a completely random search of parameters. Future exploration of other shapes and boundary conditions is planned, as well as the extension of the prediction in the non-linear deformation regime.

Acknowledgments. The authors wish to acknowledge CSC – IT Center for Science, Finland, for computational resources.

References

1. Agrawal, A., Choudhary, A.: Perspective: materials informatics and big data: Realization of the "fourth paradigm" of science in materials science. Apl. Mater. **4**(5), 053208 (2016)
2. Akusok, A., Leal, L.E., Björk, K.M., Lendasse, A.: Scikit-elm: an extreme learning machine toolbox for dynamic and scalable learning. In: International Conference on Extreme Learning Machine, pp. 69–78. Springer (2019)

3. Espinosa-Leal, L., Akusok, A., Lendasse, A., Björk, K.M.: Extreme learning machines for signature verification. In: International Conference on Extreme Learning Machine, pp. 31–40. Springer (2019)
4. Espinosa-Leal, L., Akusok, A., Lendasse, A., Björk, K.M.: Website classification from webpage renders. In: International Conference on Extreme Learning Machine, pp. 41–50. Springer (2019)
5. Gebrehiwot, S.Z., Leal, L.E., Eickhoff, J., Rechenberg, L.: The influence of stiffener geometry on flexural properties of 3d printed polylactic acid (pla) beams. In: Progress in Additive Manufacturing, pp. 1–11 (2020)
6. Leal, L.E., Björk, K.M., Lendasse, A., Akusok, A.: A web page classifier library based on random image content analysis using deep learning. In: Proceedings of the 11th PErvasive Technologies Related to Assistive Environments Conference, pp. 13–16 (2018)
7. Lu, L., Dao, M., Kumar, P., Ramamurty, U., Karniadakis, G.E., Suresh, S.: Extraction of mechanical properties of materials through deep learning from instrumented indentation. Proc. Nat. Acad. Sci. **117**(13), 7052–7062 (2020)
8. March, H.: The deflection of a rectangular plate fixed at the edges. Trans. Am. Math. Soc. **27**(3), 307–317 (1925)
9. Martinez, P., Al-Hussein, M., Ahmad, R.: A scientometric analysis and critical review of computer vision applications for construction. Auto. Constr. **107**, 102947 (2019)
10. Murakami, Y.: Theory of elasticity and stress concentration, John Wiley & Sons (2016)
11. Nie, Z., Jiang, H., Kara, L.B.: Stress field prediction in cantilevered structures using convolutional neural networks. J. Comput. Inf. Sci. Eng. **20**(1) (2020)
12. Pedregosa, F., Varoquaux, G., Gramfort, A., Michel, V., Thirion, B., Grisel, O., Blondel, M., Prettenhofer, P., Weiss, R., Dubourg, V., Vanderplas, J., Passos, A., Cournapeau, D., Brucher, M., Perrot, M., Duchesnay, E.: Scikit-learn: machine learning in python. J. Mach. Learn. Res. **12**, 2825–2830 (2011)
13. Ramakrishna, S., Zhang, T.Y., Lu, W.C., Qian, Q., Low, J.S.C., Yune, J.H.R., Tan, D.Z.L., Bressan, S., Sanvito, S., Kalidindi, S.R.: Materials informatics. J. Intell. Manuf. **30**(6), 2307–2326 (2019)
14. Słoński, M., Tekieli, M.: 2d digital image correlation and region-based convolutional neural network in monitoring and evaluation of surface cracks in concrete structural elements. Materials **13**(16), 3527 (2020)
15. Sony, S., Dunphy, K., Sadhu, A., Capretz, M.: A systematic review of convolutional neural network-based structural condition assessment techniques. Eng. Struct. **226**, 111347 (2020)
16. Spencer, B.F., Jr., Hoskere, V., Narazaki, Y.: Advances in computer vision-based civil infrastructure inspection and monitoring. Engineering **5**(2), 199–222 (2019)
17. Timoschenko, S.: Cvi. the approximate solution of two-dimensional problems in elasticity. Lond. Edinb. Dublin Philos. Mag. J. Sci. **47**(282), 1095–1104 (1924)
18. Timoshenko, S.P., Woinowsky-Krieger, S.: Theory of plates and shells. McGraw-hill (1959)
19. Yang, C., Kim, Y., Ryu, S., Gu, G.X.: Prediction of composite microstructure stress-strain curves using convolutional neural networks. Mater. Des. **189**, 108509 (2020)

On the Intensive Care Unit Admission During the COVID-19 Pandemic in the Region of Lleida, Spain: A Machine Learning Study

Didac Florensa[1], Jordi Mateo[1], Francesc Solsona[1], Pere Godoy[2], and Leonardo Espinosa-Leal[3](✉)

[1] Department of Computer Engineering, University of Lleida, C/ de Jaume II, 25001 Lleida, Spain
{didac.florensa,jordi.mateo,francesc.solsona}@udl.cat

[2] Institue of Biomedical Research of Lleida, Av. Alcalde Rovira Roure 80, 25198 Lleida, Spain
pere.godoy@gencat.cat

[3] Department of Business Management and Analytics, Arcada University of Applied Sciences, Jan-Magnus Janssons plats 1, 00560 Helsinki, Finland
leonardo.espinosaleal@arcada.fi

Abstract. The coronavirus disease 2019 (COVID-19) is an infectious and high transmissible disease that may cause severe illness. Some severe cases might require intensive care unit (ICU) admission because, in some cases, patients had comorbidities and several previous symptoms that can worst their condition. Therefore, it is of paramount importance to find strategies for helping manage the occupancy of ICU units. In this paper, we present a predictive model about ICU admission. We trained different machine learning (ML) models using a dataset containing previous symptoms and risk factors, such as demographics. The algorithms used were Random Forest (RF), Logistic Regression (LR), and Extreme Learning Machine (ELM), and the metrics used to evaluate them were accuracy, balanced accuracy, sensitivity, and specificity. We used RandomOverSampling (ROS), NearMiss (NM), and SMOTE algorithms to balance the dataset at different proportions. The best RF obtained model got a specificity of around 97%. The best LR model gives specificity about 93% and, the best ELM obtained a specificity of 94%. These results demonstrate the excellent performance of these algorithms in these kinds of datasets. Moreover, our findings show that ROS and SMOTE performed better than NM.

Keywords: SARS-CoV-2 · COVID-19 · ICU admission · Symptoms · Machine learning · Predictive models

1 Introduction

The coronavirus disease 2019 (COVID-19) is an infectious disease that may cause several cardiovascular, renal or neurological illnesses, even though the most com-

K.-M. Björk (Ed.): ELM 2021, PALO 16, pp. 92–103, 2023.
https://doi.org/10.1007/978-3-031-21678-7_9

mon is respiratory. The number of cases had increased rapidly, but information on the clinical characteristics of affected patients was limited. Furthermore, some of these patients required intensive care unit (ICU) admission [31].

The number of ICU admissions increased exponentially due to the COVID-19, and it obligated the hospital managers to adjust the managing of ICU [28]. Accurate observations found that in some COVID-19 cases, the virus affected patients heavily, even without previous risk factors but in general, patients with comorbidities presented a higher risk [5,20,25]. These studies demonstrated that risk factors as obesity, diabetes and cardiovascular disease could increase the risk of death. Therefore, risk factors could be determinant to require ICU and may be considered to build a predictive model.

Artificial intelligence is continuously improving and helping to automate the process and take decisions in industrial, medical and business services [3,13]. Machine learning (ML) can build predictive models that significantly reduce the time to understand and analyse data. Many machine learning algorithms can create models for classification or regression tasks, depending on the searched solution. In this work, we focused on the classification types. We used Random Forest, Logistic regression and Extreme Learning Machine algorithms to implement prediction models that calculate the risk of ICU admission caused by the COVID-19 disease. To implement that, we used previous symptoms related to this illness and associated risk factors.

This study's main contributions are building a model to predict ICU admission and comparing some machine learning algorithms to analyse their performance.

1.1 Related Works

Several studies have been published about machine learning algorithms used as predictive models in the health care ambit. The exponential increase in artificial intelligence in health motivates to investigate new methods to build processes and systems that help experts make decisions. For instance, one of the central areas investigated is cancer. Previous research used machine learning to build models that predict the risk of suffering breast cancer [24]. Here, the authors used risk factors to train the model in a similar way to a study done by Zhang *et al.* [33]. Here, the risk factors were taken into account to predict the risk of pancreatic cancer. Other works also applied machine learning to predict the risk of new cancer [14,30].

COVID-19 provoked a revolution in some aspects of our lives, but the most important was the health sector. This virus implied to create a new view of health ambit, specifically the importance of public health. The research community started to investigate the effect of COVID-19, new ways to stop the transmission, new medicines to decrease the risk of complications and new vaccinations. Also, they started to study why some cases can finish with more complications than others. Risk factors also were the key in some patients to understand the ICU admission and the deaths. Some studies demonstrated these in several cases [9]. The authors corroborated that risk factors as obesity, cancer or respiratory disease increase the risk of having complications. Also, some other studies demonstrated that diabetes, hypertension and metabolic syndrome were independent risk factors of death among COVID-19 patients.

These previous studies motivated us to investigate new predictive models that anticipate the virus and manage the available resources, such as the ICU. One of the problems due to the COVID-19 pandemic was the ICU collapse, and more importantly, the difficulty of being available for all the patients. Some previous studies predict the risk of ICU admission by risk factors [16,34], which demonstrated the usability of some models. Finally, other studies used machine learning algorithms to build predictive models, as we did in this work. Subudhi *et al.* [26] presented a study that used these kinds of algorithms for predicting ICU admission and mortality. Previously, another study about ICU transfer prediction in hospitalised Covid-19 patients was carried out [7].

2 Research Methodology

2.1 Dataset

The collected data correspond to new cases in the Lleida region (Spain) from March 14, 2020, until July 31, 2021. Data were collected by public health officials from the department of health by telephone interview with each patient and checking and/or reviewing the associated clinical records, as regular data collection by the public surveillance epidemiological system.

Lleida is the largest province in Catalonia (Spain), with a population density of 36 people per square kilometre. More specifically, the population was 438,001 in 2014 [1], 221,891 men and 216,110 women. The population of the Lleida region presents lifestyles, risk factors and work activity which can be linked to a specific incidence of certain types of illness. Nearly half of the population of Lleida province live in rural areas. Consequently, their lifestyle is different from that of the more urban populations in other Catalan provinces. A peculiarity of this region is the work environment. In rural areas, the main activity is the agri-food industry and, in urban areas, it is service sector activities such as education, health and catering [11].

The overall data set consisted of 56,472 SARS-CoV-2 cases confirmed by PCR diagnoses in the province of Lleida. Patients with ICU admission were 245 (0,4%) against the 56,227 (96.6%) that they did not require. This represented an imbalanced dataset, making it challenging to build a predictive model that predicts ICU admission. To solve this problem, we used different kinds of sampling techniques that balance data.

More specifically, the training data set consisted of 80% of the total cases (6,262), and the test was the remaining 20% (1,565). Each corpus register contains the following variables: age group (0–14, 15–34, 35–44, 45–54, 55–64, 65–74, 75–84, ≥85); gender (male, female); fever (Yes, No); cough (Yes, No); pneumonia (Yes, No); sore throat (Yes, No); shivers (Yes, No); dyspnea (Yes, No); vomiting (Yes, No); diarrhoea (Yes, No); respiratory distress syndrome (Yes, No); acute renal failure (Yes, No); diabetes (Yes, No); cardiovascular disease (Yes, No);

chronic lung disease (Yes, No); neurological disease (Yes, No); hospitalised (Yes, No); asymptomatic (Yes, No); ICU (Yes, No) Fig. 1.

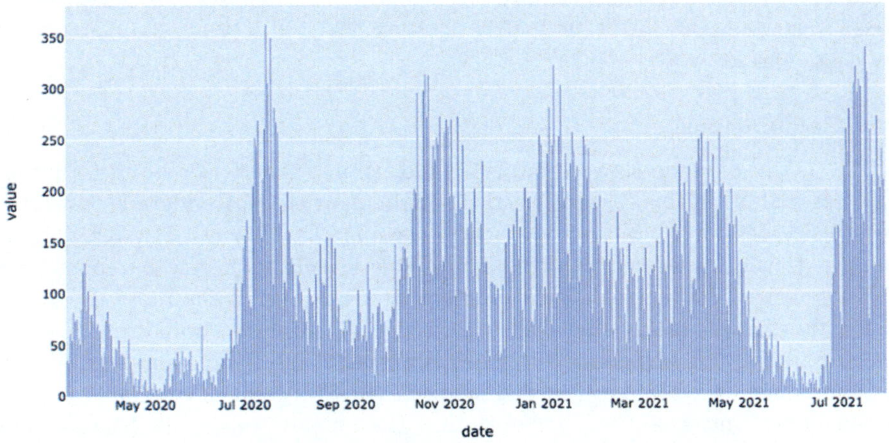

Fig. 1. Evolution of the epidemiological curve in Lleida region (Spain). Cases from March 14, 2020 until July 31, 2021. In the figure, the x-axis represents the day of diagnoses and the y-axis the number of cases.

2.2 Ethical Considerations

All data were anonymised before the study to protect patient privacy and confidentiality. This study was part of the public health response to coronavirus outbreaks. This project was approved by the reference ethics committee (Committee of Ethics and Clinical Research of Lleida - CEIC).

2.3 Preprocessing

Classification algorithms applied to imbalanced datasets have serious weaknesses, since good global accuracy does not implicitly good performance for minority class [23]. The most common solution to deal with imbalanced datasets is the use of sampling algorithms [18]. These sampling algorithms are divided into two types of techniques: under and oversampling. The undersampling method balances the dataset by keeping all the data in the minority class and decreasing the size of the majority class [27]. On the other hand, oversampling technique keeps all of the data in the majority class and increases the size of the minority class [4]. In conclusion, these algorithms aim to balance the dataset to obtain betters performance results.

The Synthetic Minority Over-sampling Technique (SMOTE) [6] is the most common technique used in scientific research. This algorithm focus on creating synthetic instances of the minority class. This method has been used successfully in some cancer research areas [12,32]. RandomOverSampler is a similar algorithm because it increases the number of cases of the minority class, but it generates the information randomly [21]. The difference between traditional methods and SMOTE is that the latter only over-samples the borderline examples of the minority class [15].

2.4 Calculations

All the predictive models were implemented using a supervised artificial intelligence algorithm. In this case, we used the main algorithms: Random Forest (RF), Logistic Regression (LR) and Extreme Learning Machine (ELM). RF consists of many decision trees that it utilises ensemble learning [8]. It was implemented with a forest of 300 decision tree predictors. LR is one of the most used algorithms for classification. It is based on a sigmoid function and works best on binary classification problems [19]. ELM is a recent algorithm composed of a single hidden layer feed-forward neural network (SLFNs), which randomly chooses hidden nodes and analytically determines the output weights of SLFNs. This algorithm tends to provide good generalisation performance at extremely fast learning speed for different type of problems [10,17].

There are several metrics to evaluate classification algorithms. In this case, the metrics were the accuracy, the specificity, the sensitivity and the balanced accuracy. Analysing such metrics and their significance must be interpreted correctly to evaluate these algorithms and their performance. Accuracy represents the result of a measurement, calculation, or specification that conforms to the correct value or a standard. In our case, it represents the ratio between the correctly classified samples. The sensitivity represents the proportion of correctly classified negative (TN) samples compared to the total number of negative samples. The specificity is the proportion of correctly classified positive (TP) samples compared to the total number of positive samples. A False positive (FP) is when the model produces a positive outcome, and it was initially negative. The opposite case is called false negative (FN). Finally, the balanced accuracy represents true positive and true negative predictions by the number of positive and negative samples, respectively, dividing their sum by two. That is, it is the average between Sensitivity and Specificity [29].

The models in this work were implemented using the Python programming language, and the calculations were performed using scikit-learn [22] and scikit-ELM [2] python libraries.

- **Accuracy**: Ratio between the correctly classified samples
$$\frac{TP+TN}{TP+TN+FP+FN}.$$
- **Sensitivity**: Proportion of correctly classified positive samples compared to the total number of positive samples.
$$\frac{TP}{TP+FN}.$$

- **Specificity**: Proportion of correctly classified negative samples compared to the total number of negative samples.

$$\frac{TN}{TN + FP}.$$

- **Balanced Accuracy**: The average between Sensitivity and Specificity.

$$\frac{Sensitivity + Specificity}{2}.$$

3 Results and Discussion

This section shows and discusses the performance obtained by different machine learning algorithms and by different sampling algorithms. Table 1 shows the outcomes obtained by RF, LR and ELM and by RandomOverSampling, NearMiss and SMOTE sampling algorithm. First of all, we present the metrics obtained by the RandomOverSampling. RF produces a bad performance with the original imbalanced dataset. The accuracy was 99.5%, the sensitivity was 99.9%, but the specificity was 9.4%. In consequence, the balanced accuracy was 54.6% which suggested this model is inefficient. However, with the 20% of sampling in the minority class, the performance model increased notably. The accuracy was 98.0%, the sensitivity was 98.4%, the specificity was 96.3%, and the balanced accuracy was 97.3%. These results indicated the improvement of the performance model with few synthetic cases.

Moreover, similar outcomes were obtained when the model was trained with 50% and 100% (balanced dataset) of sampling. The accuracy was 97.3%, the sensitivity was 97.5% and the specificity 97.0% with the completely balanced dataset. These few differences suggested that with fewer synthetic cases, it could obtain an excellent model. Also, these results suggested that this model could consider it excellent [7]. LR also obtained similar performance while the sampling was increasing. The obtained metrics with the imbalanced dataset were next: accuracy of 99.5%, the sensitivity of 100% and specificity of 0%.

On the other hand, with the 20% of sampling, the performance improves significantly. The accuracy was 95.8%, the sensitivity 94.4% and the specificity 92.5%. Furthermore, with the wholly balanced dataset, the metrics kept around 95%, as the Table 1 shows. Finally, the ELM also got similar metrics to RF and LR. The accuracy with the initial dataset (imbalanced) was 99.6%, the sensitivity was 100%, and the specificity was 0%, but with a 20% of sampling, the metrics increase positively. The improved accuracy was 95.8%, the sensitivity was 96.4%, and the specificity was 92.9%. The metrics kept similar as these once when the sampling was increased.

Next, we present the obtained metrics by the NearMiss sampling algorithm. In this case, this sampling method did not keep the excellent performance obtained previously in the Table 1. RF obtained an accuracy of 86.4%, a sensitivity of 94.5% and specificity of 70.2% with a 50% of sampling. The specificity did not increase as RandomOverSampling increased before. However, with the

Table 1. Comparison of the performance metrics for three different machine learning algorithms: Random Forest (RF), Logistic Regression (LR) and Extreme Learning Machine (ELM). The obtained values without data imbalance strategy (none) and three different sampling methods: RandomOverSampling (ROS), NearMiss sampling (NM) and SMOTE (Synthetic Minority Over-sampling Technique) algorithms.

ML algorithm	Imbalanced Method	Sampling	Accuracy	Sensitivity	Specificity
RF	-	-	99.5	99.9	9.4
	ROS	20%	98.0	98.4	96.3
		50%	97.6	98.2	96.4
		100%	97.3	97.5	97.0
	NM	20%	89.8	95.6	60.8
		50%	86.4	94.5	70.2
		100%	91.8	100.0	83.7
	SMOTE	20%	97.3	98.3	92.7
		50%	96.3	97.9	92.9
		100%	95.7	96.8	94.8
LR	-	-	99.5	100	0
	ROS	20%	95.8	94.4	92.5
		50%	95.2	96.4	92.8
		100%	94.6	96.4	92.7
	NM	20%	90.0	97.0	55.4
		50%	81.0	91.8	59.4
		100%	81.6	86.3	77.0
	SMOTE	20%	95.6	96.8	89.7
		50%	94.7	96.5	91.1
		100%	94.1	96.4	91.8
ELM	-	-	99.6	100.0	0.0
	ROS	20%	95.8	96.4	92.9
		50%	95.3	96.4	93.1
		100%	95.4	96.3	94.4
	NM	20%	89.1	98.4	43.2
		50%	83.3	98.0	54.1
		100%	84.4	91.8	77.0
	SMOTE	20%	95.5	96.4	91.0
		50%	95.0	96.8	91.5
		100%	94.2	96.6	91.8

100% of sampling, the model improved the performance notably. The accuracy was 91.8%, the sensitivity 100% and the specificity 83.7%. Even though the specificity was not around the 90%, it can consider a good predictive model.

LR obtained an accuracy of 81%, a sensitivity of 91.8%, but the specificity was 59.4% with the 50% of sampling. In this case, the model could not consider a good option because the obtained specificity was under 60%. Moreover, with the 100% of sampling, the specificity improved until 77%, but the other metrics kept or decreased than previous training. The accuracy was 81.6%, and the sensitivity was 86.3%. In this case, analysing the obtained metrics, the model could be considered a good model, but other algorithms can improve these metrics as RF showed. Finally, ELM trends were similar to LR. The accuracy with the 20% of sampling was 89.1%, the sensitivity was 98.4%, but the specificity was 43.2%. Similar results were obtained with the 50% of sampling, but the total improvement was obtained with 100% of sampling. The accuracy was 84.4%, the sensitivity was 91.8%, and the specificity was 77%.

Finally, we show the results obtained by the SMOTE sampling algorithm. The built models got similar metrics as RandomOverSampling obtained. RF also was the first algorithm trained. The accuracy obtained with the 50% of sampling was 96.3%, the sensitivity was 97.9%, and the specificity was 92.9%. These results suggested that the model could consider it excellent without the completely balanced dataset. The 100% of sampling corroborated the good performance. The metrics also kept around the 95%, as previous training obtained. LR also got similar results to RF. The accuracy was 94.7%, the sensitivity was 96.5%, and the specificity was 91.1% with the 50% of sampling. Finally, with the 100% of sampling (balanced dataset), the model also obtained similar results, changing only the decimals in the performance. This small variation also suggested that it was unnecessary to alter the dataset with synthetic data to obtain excellent models. Finally, the ELM obtained similar trends as the presented previous algorithms. The metrics improved only with the 20% of sampling, and they also kept while the sampling was increasing. The accuracy with 20% of sampling was 95.5%, the sensitivity was 96.4%, and the specificity was 91.0%.

In general, the RandomOverSampling and Smote obtained excellent performance in the trained algorithms (RF, LR and ELM) with few synthetic cases. These algorithms obtained metrics over 90% with only 20% of sampling. This fact suggested that the model could be considered excellent without altering so much the initial dataset. Also, permit to demonstrate that the ELM obtained similar outcomes to RF and LR. Moreover, they also conclude that these algorithms could be used to build a predictive model for these specific datasets. By cons, we found the NearMiss sampling because the ML training with it could not obtain similar performance, even though they were not bad. In RF training, they could obtain metrics over 80%.

4 Conclusion and Future Research

The COVID-19 stimulated some research about the use of machine learning algorithms to explore the possibility of predicting this illness's evolution. Previous studies focused on trying to predict the risk of ICU admission by previous comorbidities and symptoms. However, in this study, we used random forest, logistic regression and extreme machine learning as algorithms to build the most efficient predictive model and compare their performance.

Some essential outcomes were found, such as using oversampling algorithms to balance the dataset and using some ML algorithms. SMOTE and RandomOverSampling obtained a good performance with few synthetic cases. NearMiss needed to balance the classes completely to obtain significant metrics. The ML algorithms' metrics were similar in all tests; therefore, they could be considered an excellent algorithm to build a predictive model. Also, in some cases, the metrics were over 95%, which suggested the quality of the model.

Regarding future research, it is essential to highlight the importance of each feature in the dataset and try to quantify the risk depending on these features. Also, enlarge the dataset with the full days in the ICU unit to predict its time. Furthermore, train different ML algorithms to find the best predictive model.

Acknowledgments. The authors wish to acknowledge CSC – IT Center for Science, Finland, for computational resources.

References

1. Idescat. Anuari estadístic de Catalunya. Densitat de població. Comarques i Aran, àmbits i províncies (2014). https://www.idescat.cat/pub/?id=aec&n=249&t=2014
2. Akusok, A., Espinosa Leal, L., Björk, K.M., Lendasse, A.: Scikit-elm: an extreme learning machine toolbox for dynamic and scalable learning. In: International Conference on Extreme Learning Machine, pp. 69–78. Springer (2019)
3. Ao, Y., Li, H., Zhu, L., Ali, S., Yang, Z.: The linear random forest algorithm and its advantages in machine learning assisted logging regression modeling. J. Pet. Sci. Eng. **174**, 776–789 (Mar 2019). https://doi.org/10.1016/J.PETROL.2018.11.067
4. Barua, S., Islam, M.M., Yao, X., Murase, K.: MWMOTE - Majority weighted minority oversampling technique for imbalanced data set learning. IEEE Trans. Knowl. Data Eng. **26**(2), 405–425 (Feb 2014). https://doi.org/10.1109/TKDE.2012.232
5. Buetti, N., Ruckly, S., de Montmollin, E., Reignier, J., Terzi, N., Cohen, Y., Siami, S., Dupuis, C., Timsit, J.F.: COVID-19 increased the risk of ICU-acquired bloodstream infections: a case-cohort study from the multicentric OUTCOMEREA network. Intensive Care Med. **47**(2), 180–187 (Jan 2021). https://doi.org/10.1007/S00134-021-06346-W, https://link.springer.com/article/10.1007/s00134-021-06346-w
6. Chawla, N.V., Bowyer, K.W., Hall, L.O., Kegelmeyer, W.P.: SMOTE: Synthetic minority over-sampling technique. J. Artif. Intell. Res. **16**, 321–357 (Jun 2002). https://doi.org/10.1613/jair.953, https://www.jair.org/index.php/jair/article/view/10302
7. Cheng, F.Y., Joshi, H., Tandon, P., Freeman, R., Reich, D.L., Mazumdar, M., Kohli-Seth, R., Levin, M.A., Timsina, P., Kia, A.: Using machine learning to predict ICU transfer in hospitalized COVID-19 patients. J. Clin. Med. **9**(6), 1668 (Jun 2020). https://doi.org/10.3390/JCM9061668, https://www.mdpi.com/2077-0383/9/6/1668/htmhttps://www.mdpi.com/2077-0383/9/6/1668

8. Cutler, A., Cutler, D.R., Stevens, J.R.: Random forests. Springer, Boston, MA, Boston, MA (2012). https://link.springer.com/chapter/10.1007/978-1-4419-9326-7_5
9. Esai Selvan, M.: Risk factors for death from COVID-19. Nat. Rev. Immun. **20**(7), 407–407 (May 2020). https://doi.org/10.1038/s41577-020-0351-0, https://www.nature.com/articles/s41577-020-0351-0
10. Espinosa-Leal, L., Akusok, A., Lendasse, A., Björk, K.M.: Website classification from webpage renders. In: International Conference on Extreme Learning Machine, pp. 41–50. Springer (2019)
11. Florensa, D., Godoy, P., Mateo, J., Solsona, F., Pedrol, T., Mesas, M., Pinol, R.: The use of multiple correspondence analysis to explore associations between categories of qualitative variables and cancer incidence. IEEE J. Biomed. Health Inf. **25**(9), 3659–3667 (Sep 2021). https://doi.org/10.1109/JBHI.2021.3073605
12. Geetha, R., Sivasubramanian, S., Kaliappan, M., Vimal, S., Annamalai, S.: Cervical cancer identification with synthetic minority oversampling technique and PCA analysis using random forest classifier. J. Med. Syst. **43**(9), 1–19 (2019). https://doi.org/10.1007/s10916-019-1402-6
13. Gianfrancesco, M.A., Tamang, S., Yazdany, J., Schmajuk, G.: Potential biases in machine learning algorithms using electronic health record data. JAMA Intern. Med. **178**(11), 1544–1547 (Nov 2018). https://doi.org/10.1001/JAMAINTERNMED.2018.3763, https://jamanetwork.com/journals/jamainternalmedicine/fullarticle/2697394
14. H, B., JM, H., M, T., VM, K., A, M.: Predicting breast cancer risk using interacting genetic and demographic factors and machine learning. Sci. Rep. **10**(1) (Dec 2020). https://doi.org/10.1038/S41598-020-66907-9, https://pubmed.ncbi.nlm.nih.gov/32632202/
15. Han, H., Wang, W.Y., Mao, B.H.: Borderline-SMOTE: a new over-sampling method in imbalanced data sets learning. In: Lecture Notes in Computer Science (including subseries Lecture Notes in Artificial Intelligence and Lecture Notes in Bioinformatics) LNCS. vol. 3644, pp. 878–887. Springer (2005). https://link.springer.com/chapter/10.1007/11538059_91
16. He, F., Quan, Y., Lei, M., Liu, R., Qin, S., Zeng, J., Zhao, Z., Yu, N., Yang, L., Cao, J.: Clinical features and risk factors for ICU admission in COVID-19 patients with cardiovascular diseases. Aging Dis. **11**(4), 763 (2020). https://doi.org/10.14336/AD.2020.0622, /pmc/articles/PMC7390529/ /pmc/articles/PMC7390529/?report=abstract https://www.ncbi.nlm.nih.gov/pmc/articles/PMC7390529/
17. Huang, G.B., Zhu, Q.Y., Siew, C.K.: Extreme learning machine: theory and applications. Neurocomputing **70**(1–3), 489–501 (Dec 2006). https://doi.org/10.1016/J.NEUCOM.2005.12.126
18. KaurHarsurinder, Singh, P., Kaur, M.: A Systematic review on imbalanced data challenges in machine learning. ACM Comput. Surv. (CSUR) **52**(4) (Aug 2019). https://doi.org/10.1145/3343440, https://dl.acm.org/doi/abs/10.1145/3343440
19. Lemon, S.C., Roy, J., Clark, M.A., Friedmann, P.D., Rakowski, W.: Classification and regression tree analysis in public health: methodological review and comparison with logistic regression. Ann. Behav. Med. **26**(3), 172–181 (2003). https://link.springer.com/article/10.1207/S15324796ABM2603_02
20. Lighter, J., Phillips, M., Hochman, S., Sterling, S., Johnson, D., Francois, F., Stachel, A.: Obesity in patients younger than 60 years is a risk factor for COVID-19 hospital admission. Clin. Infect. Dis. **71**(15), 896–897

(Jul 2020). https://doi.org/10.1093/CID/CIAA415, https://academic.oup.com/cid/article/71/15/896/5818333
21. Lu, W., Hou, H., Chu, J.: Feature fusion for imbalanced ECG data analysis. Biomed. Sig. Process. Control **41**, 152–160 (Mar 2018). https://doi.org/10.1016/J.BSPC.2017.11.010
22. Pedregosa, F., Varoquaux, G., Gramfort, A., Michel, V., Thirion, B., Grisel, O., Blondel, M., Prettenhofer, P., Weiss, R., Dubourg, V., Vanderplas, J., Passos, A., Cournapeau, D., Brucher, M., Perrot, M., Duchesnay, E.: Scikit-learn: machine learning in python. J. Mach. Learn. Res. **12**, 2825–2830 (2011)
23. Rahman, M.M., Davis, D.N.: Addressing the class imbalance problem in medical datasets. Int. J. Mach. Learn. Comput. 224–228 (2013). https://doi.org/10.7763/IJMLC.2013.V3.307
24. RD, N., T, A., L, L., I, D.: Diagnostic accuracy of different machine learning algorithms for breast cancer risk calculation: a meta-analysis. Asian Pac. J. Cancer Prev. : APJCP **19**(7), 1747–1752 (Jul 2018). https://doi.org/10.22034/APJCP.2018.19.7.1747, https://pubmed.ncbi.nlm.nih.gov/30049182/
25. Roncon, L., Zuin, M., Rigatelli, G., Zuliani, G.: Diabetic patients with COVID-19 infection are at higher risk of ICU admission and poor short-term outcome. J. Clin. Virol. **127**, 104354 (Jun 2020). https://doi.org/10.1016/J.JCV.2020.104354
26. Subudhi, S., Verma, A., Patel, A.B., Hardin, C.C., Khandekar, M.J., Lee, H., McEvoy, D., Stylianopoulos, T., Munn, L.L., Dutta, S., Jain, R.K.: Comparing machine learning algorithms for predicting ICU admission and mortality in COVID-19. npj Digit. Med. **4**(1), 1–7 (May 2021). https://doi.org/10.1038/s41746-021-00456-x, https://www.nature.com/articles/s41746-021-00456-x
27. Sun, Z., Song, Q., Zhu, X., Sun, H., Xu, B., Zhou, Y.: A novel ensemble method for classifying imbalanced data. Pattern Recogn. **48**(5), 1623–1637 (May 2015). https://doi.org/10.1016/J.PATCOG.2014.11.014
28. Tartari, F., Guglielmo, A., Fuligni, F., Pileri, A.: Changes in emergency service access after spread of COVID19 across Italy. J. Eur. Acad. Dermatol. Venereology **34**(8), e350–e351 (Aug 2020). https://doi.org/10.1111/JDV.16553,/pmc/articles/PMC7267617/?report=abstract https://www.ncbi.nlm.nih.gov/pmc/articles/PMC7267617/
29. Tharwat, A.: Classification assessment methods. Appl. Comput. Inf. **17**(1), 168–192 (2020)
30. Ting, W.C., Lu, Y.C.A., Ho, W.C., Cheewakriangkrai, C., Chang, H.R., Lin, C.L.: Machine learning in prediction of second primary cancer and recurrence in colorectal cancer. Int. J. Med. Sci. **17**(3), 280–291 (2020). https://doi.org/10.7150/IJMS.37134
31. Wang, D., Hu, B., Hu, C., Zhu, F., Liu, X., Zhang, J., et. al: Clinical characteristics of 138 hospitalized patients with 2019 novel coronavirus-infected pneumonia in Wuhan, China. JAMA **323**(11), 1061–1069 (Mar 2020). https://doi.org/10.1001/JAMA.2020.1585, https://jamanetwork.com/journals/jama/fullarticle/2761044
32. Yan, S., Qian, W., Guan, Y., Zheng, B.: Improving lung cancer prognosis assessment by incorporating synthetic minority oversampling technique and score fusion method. Med. Phys. **43**(6Part1), 2694–2703 (May 2016). https://doi.org/10.1118/1.4948499, http://doi.wiley.com/10.1118/1.4948499
33. Zhang, Y., Zhu, S., Yuan, Z., Li, Q., Ding, R., Bao, X., Zhen, T., Fu, Z., Fu, H., Xing, K., Yuan, H., Chen, T.: Risk factors and socio-economic burden in pancreatic ductal adenocarcinoma operation: a machine learning based analysis. BMC Cancer **20**(1), 1–12 (Nov 2020). https://doi.org/10.1186/S12885-020-07626-2, https://bmccancer.biomedcentral.com/articles/10.1186/s12885-020-07626-2

34. Zhao, Z., Chen, A., Hou, W., Graham, J.M., Li, H., Richman, P.S., Thode, H.C., Singer, A.J., Duong, T.Q.: Prediction model and risk scores of ICU admission and mortality in COVID-19. PLOS ONE **15**(7), e0236618 (Jul 2020). https://doi.org/10.1371/JOURNAL.PONE.0236618, https://journals.plos.org/plosone/article?id=10.1371/journal.pone.0236618

Verification of Static Signatures using Dynamic Time Warping on Features from High Pressure Points

Ruben Acosta-Velasquez[1], Leonardo Espinosa-Leal[2(✉)], Alexander Garcia-Perez[1], and Kaj-Mikael Björk[2,3]

[1] Universidad EAN, Bogota, Colombia
{rdacosta,agarciap}@universidadean.edu.co
[2] Department of Business Management and Analytics, Arcada University of Applied Sciences, Jan-Magnus Janssons plats 1, 00560 Helsinki, Finland
{leonardo.espinosaleal,kaj-mikael.bjork}@arcada.fi
[3] Hanken School of Economics, Arkadiagatan 22, 00100 Helsinki, Finland

Abstract. In this paper, we present a novel methodology for the classification of static handwritten signatures. We extract features from each user's high-pressure points by filtering the skeletonized-binarized digital signatures at seven different thresholds. We compare the sequences obtained using the Dynamic Time Warping (DTW) algorithm with five metrics for the x and y axes. The resulting distances are used as features in a binary dataset of genuine-genuine and genuine-forgery signatures with fourteen features. We train and compare three classifiers: Random Forest (RF), extreme gradient boosting (XGBoost), and Extreme Learning Machines (ELM) on the processed data from the MCYT-75 dataset [12]. Our results show that all models perform with similar accuracy, where the Euclidean and Manhattan metrics perform the best.

Keywords: Signature verification · Dynamic Time Warping · Deep learning · Extreme learning machines

1 Introduction

The current technological advancements have allowed a large part of our information systematization: ID numbers, medical records, web content, bank accounts, and others. Because of this, identity verification systems have become important regarding security. These systems mainly use sophisticated models based on recognizing patterns from biometric quantities such as fingerprints [9,13], retina, signatures [11] and other face features [4].

There are two types of biometric systems: static, related to physical features, and dynamics, related to behaviour patterns. The most common static features that stand out are fingerprints, hands' geometry, the shape of the face, retina, Etc. The most common dynamics features are the voice, facial gestures, walking, handwriting, among others.

K.-M. Björk (Ed.): ELM 2021, PALO 16, pp. 104–113, 2023.
https://doi.org/10.1007/978-3-031-21678-7_10

An important area of research deals with the problem of handwritten recognition. Handwritten signature recognition is essential for validating bank cheques, documents, and transactions. A handwritten signature can be seen as a stroke or image done by a person on a document to confer validity or seal an agreement. Automatic systems designed for recognition are divided into online and offline systems; the first one requires the user's physical presence and a device that aids with digitization. The second one does not require the user's physical presence; however, it requires extracting features that it takes from a physical document. Online systems have some advantages, such as their ability to capture features that offline systems are not capable of, such as the stroke's velocity and the user pressure when the user signs.

The primary use of these devices responds to the possibility of impersonation fraud. Genuine signatures, which refer to original signatures registered by the user, can be forged by a different user. Forgeries are divided into three kinds: skilled forgeries, which are produced when the forger has unrestricted access to one or more samples of the genuine signature; casual forgeries or forgeries without training, that is when the forger has limited access to one sample within a short period, and unexpected forgeries, that is when the forger only knows the name of the user.

There are several types of verification systems for handwritten signatures. These are classified in two: *generative models* and *discriminate models*. The first ones are probabilistic for every variable involved, and the second ones involve conditional probabilities. Discriminative models require training for each genuine signature and its respective forgeries; generative models, on the other hand, require training just for a sample of genuine signatures. In this work, we study a discriminative model based on comparing the geometry of the signatures through image morphology and the DTW algorithm [16].

2 Related Works

The biometric term is directly associated with measuring a set of selected features from the human body, some of which correspond to morphological characteristics and other behavioural. These attributes are exploited for designing systems that allow verifying or identifying a specific user. For verification, it is related to the person's identity, and for identification, it is related to the necessity of validating the identity of a person previously registered in a database. Handwritten signatures have a special place in the biometrics systems. This is mainly due to its importance in common legal and administrative aspects in modern society. Different machine learning methods have been used previously for signature verification, neural networks [10,11], Markov chains, extreme learning machines [3,7], among many others.

3 Capture and Preprocessing

In developing a handwritten signature verification system, a question emerges: What is the capturing methodology of signatures? In online systems, the capture is made by digitization tablet where there exists an essential difference insomuch as the tablet allows storing features such as coordinates, pressure, inclination, and speed of the stroke. On the other hand, in offline systems, the capture is not automatic. Before any step, it is necessary to do some preprocessing. The first step is to obtain the image from a physical document and its subsequent digitization using image morphology operations. The second preprocessing step requires applied geometric transformations to set up the samples at the same scale and inclination.

3.1 Image Morphology

The capture of the signature is done using a scanner to establish it like a digital image. The first step in this process is transforming the signature in a graph with one pixel of thickness, which is reached through image morphology operations. The first step is a binarisation to transform the signature into a black and white pixels, and the second step is a skeletization function to obtain the skeleton of the signature.

Definition 1 *Let A be an image, the skeleton of A can be expressed in terms of erosion and openings, That is, it can be shown [8]*

$$S(A) = \bigcup_{k=0}^{k} S_k(A) \tag{1}$$

where,

$$S_k(A) = (A \ominus kB) - (A \ominus kB) \circ B \tag{2}$$

where B is a structuring element, and $(A \ominus kB)$ *indicates k successive erosions of A.*

3.2 Normalization

In signature comparison, it is necessary that the genuine signature and incoming signature are in the same size, position and inclination which it allows a comparison in the same "scale", in which this normalization is done by [1]:

$$S' = \begin{bmatrix} X_i \\ Y_i \end{bmatrix} = \frac{1}{D} \begin{bmatrix} P & -Q \\ Q & P \end{bmatrix} \begin{bmatrix} \bar{x}_i \\ \bar{y}_i \end{bmatrix} \tag{3}$$

$$P = \sum_{i=1}^{n} \bar{x}_i a_i + \bar{y}_i b_i \tag{4}$$

$$Q = \sum_{i=1}^{n} \bar{x}_i b_i - \bar{y}_i a_i \tag{5}$$

$$D = \sqrt{(P^2 + Q^2)\sum_{i=1}^{n}(\bar{x}_i^2 + \bar{y}_i^2)} \tag{6}$$

Where S' is the normalized signature, $\bar{x}_i = x_i - x_z$ y $\bar{y}_i = y_i - y_z$, x_i, y_i, a_i, b_i are the coordinates for the genuine and incoming signatures respectively.

4 Comparison of Signatures

Since the previous process, for the genuine signature S and an incoming signature T, four vectors of components are obtained S_x, T_x, S_y, T_y, where these are compared respectively. Practically these are defined as a $S_X = \{S_x(i)\backslash i = 1, ..., n\}$ and $T_x = \{T_x(j)\backslash j = 1, ..., m\}$ sequence where the components are dependent on i furthermore n and m are positive integers; therefore these will be compared by the DTW (Dynamic Time Warping) algorithm.

4.1 Dynamic Time Warping (DTW)

The DTW is a dynamic programming technique based on the DTW algorithm for measuring the similarity between two temporal sequences under some constraints, primarily in speed. This technique is used in comparison of voice patterns, data mining, information recovery and the detection of shape similarity for sequences in different phases.

Let $X := (x_1, x_2, ..., x_N)$ and $Y := (y_1, y_2, ..., y_M)$ be two sequences with $N, M \in \mathbb{N}$ consisting of features sequences sampled in equidistant points. There exists a feature space denoted by, $\mathbb{F}$ such that $x_n, y_n \in \mathbb{F}$ for $n \in [1 : N]$ and $m \in [1 : M]$. To compare, $x, y \in \mathbb{F}$ we required a measure of total cost defined by

$$c : \mathbb{F} \times \mathbb{F} \longrightarrow \mathbb{R}_{\geq 0} \tag{7}$$

The first step for the algorithm is to set up the cost matrix C where each matrix element performs the distance between pairs of elements from X and Y, thus:

$$C \in \mathbb{R}^{N \times M} \tag{8}$$

where $c_{ij} := d(x_i, y_j), i \in [1 : N], j \in [1 : M]$.

The optimal path is obtained from the cost matrix starting from the element in the position $(1, 1)$ to the position (N, M). The next definition formalizes the alignment notion.

Definition 2 *A* (N, M) *– path is a sequence* $p = (p_1, p_2, ..., p_L)$ *with* $p_l = (n_l, m_l) \in [1 : N] \times [1 : M]$ *for* $l \in [1 : L]$ *if*

- *Boundary condition:* $p_1 = (1, 1)$ *y* $p_L = (N, M)$

- *Monotonicity condition:* $n_1 \leq n_2 \leq ... \leq n_L$ *y* $m_1 \leq m_2 \leq ... \leq m_L$
- *Step size condition:* $p_{l+1} - p_l \in \{(1,0),(0,1),(1,1)\}$ *for* $l \in [1 : L-1]$

The total cost $c_p = (X, Y)$ of a path p between X and Y of the measure of local cost c is defined by

$$c_p(X, Y) = \sum_{l=1}^{L} c(x_{n_l}, y_{m_l}) \tag{9}$$

In addition, an optimal path between X and Y is a $p*$ path with the minimum cost of all possible paths. Then the distance, DTW, $DTW(X, Y)$ is defined as the total cost of $p*$ by

$$DTW(X, Y) := c_{p*}(X, Y) = \min\{c_p(X, Y) \backslash p \text{ is a } (N, M) - path\} \tag{10}$$

On the other hand,

$$D(n, m) := DTW(X(1 : n), Y(1 : m)) \tag{11}$$

where $D(n, m)$ is the cumulative cost matrix, whose size is $N \times M$ and $D(N, M) = DTW(X, Y)$.

Theorem 1. *The cumulative cost matrix D satisfies the identities:*

- $D(n, 1) = \sum_{k=1}^{n} c(x_k, y_1)$ *for* $n \in [1 : N]$
- $D(1, m) = \sum_{k=1}^{m} c(x_1, y_k)$ *for* $m \in [1 : M]$
- $D(n, m) = \min\{D(n-1, m-1), D(n-1, m), D(n, m-1)\} + c(x_n, y_m)$ *for* $1 < n \leq N$ *and* $1 < m \leq M$

4.2 Distance Metric

Different distances can be used for measuring the distance among the obtained sequences via DTW. In this work, we have used five different methods. In addition to the classical Euclidean metric (root sum of squared differences) and the sum of absolute differences, aka Manhattan metric; we have performed calculations using the square of the Euclidean metric (squared metric), the symmetric Kullback-Leibler metric, and the (triangular) Global Alignment Kernel (GAK) [6] metric.

5 Research Methodology

This work compares signatures using a treatment of their image morphology, focused on the pressure variable obtained from the offline signature. Typically, pressure is a feature hard to obtain from a signature in a document. This feature is easily captured via online devices. Here, we have considered a strategy where the pressure is obtained through the intensity of the pixels. With this method,

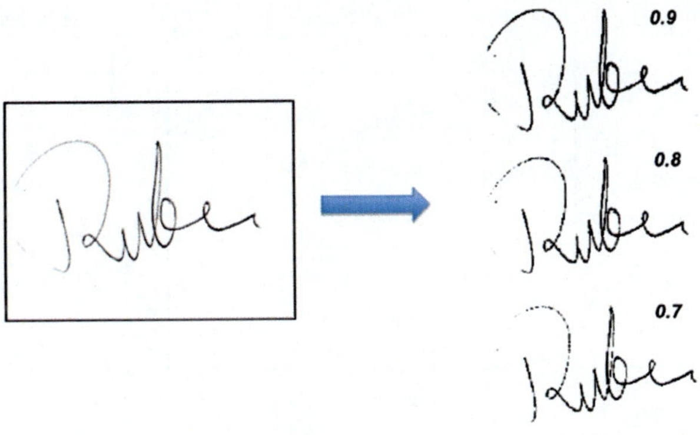

Fig. 1. (*Left*): an original scanned signature transformed (*right*) upon binarisation with three different thresholds: 0.9, 0.8 and 0.7 (from top to bottom).

the pixels are being eliminated following their intensity level to a given threshold, then some pixels with lower intensity vanish from the image [15]. Intuitively, the most intense pixels are the pixels where the pressure is the *strongest*; therefore, if the threshold increases, more pixels disappear (see Fig. 1).

For testing our method, we use the MCYT-75 dataset [12]. The dataset used in this article includes 75 users, each one with 30 signatures, distributed as 15 genuine signatures and 15 forgeries. The whole dataset of signatures is transformed following the methods described in Sect. 3. Additionally, the binarisation step is performed between the feature extraction upon digitization-skeletonization-normalization and the DTW discrimination. In this step, the significant features given by the differences in pressure and the writing brush strokes of the signature are extracted by considering filtering from the same sample to different thresholds during the binarisation step. The process works as follows: for each pair of signatures, the genuine and incoming signatures are filtered at the same threshold and compared, respectively. The image process is followed by a normalization process that sets up the signatures at the same scale and inclination.

The coordinates for x and y directions are extracted for each signature. The comparison between signatures is made between their coordinate sequences using the DTW algorithm (see Fig. 2). Each step of the methodology used in this work is described as follows:

1. The capture of signatures into the system.
2. Preprocessing, which consists of image morphology transformations, including binarisation.
3. Different thresholds applied to simulate the capture of a feature of pressure.
4. Extraction of coordinates of each signature in x and y components.
5. Operations of scaling, rotation, and translation.
6. Comparison, by considering the sample obtained for each threshold.

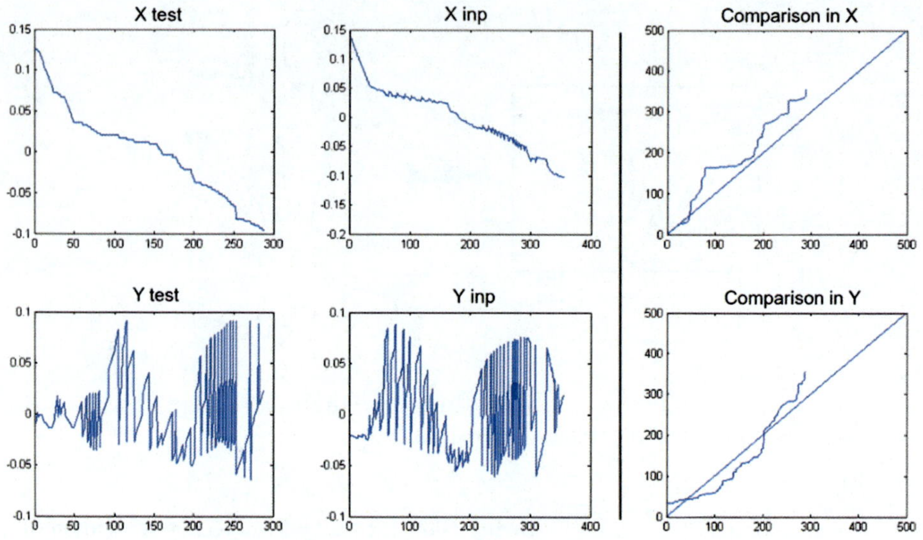

Fig. 2. Extracted sequences for both x and y directions from two different signatures. On the *right*, the respective comparison between the series via DTW.

The sequence of steps presented above describes the process of obtaining a dataset composed of two data classes: the first one, a subset whose elements are sequences of results obtained by comparing only genuine signatures. The second one is a subset whose elements are sequences of results obtained by comparing genuine signatures with forgeries. Each of these sequences contains 14 features, where seven are the results of the DTW comparison per threshold in the X component and the other seven are the DTW comparison results per threshold in the Y component. The thresholds ranged from 0.3 to 0.9.

The dataset obtained consists of 24750 rows, where 7136 samples are labelled as the true class, equivalent to the comparison between two genuine signatures, and 17614 samples are labelled as false, equivalent to the comparison between one forgery and one genuine signature. In Fig. 3 the distributions for one user at two thresholds are presented. Here we can see that the features for the genuine-genuine and forgery-genuine are well separated. We trained three different machine algorithms on this data.

6 Results and Discussion

The dataset obtained using DTW is used for training different models for the five different metrics. We have three machine learning models, two tree-based models: Random Forest (RF) and extreme gradient boosting (XGBoost), and a third model, named extreme learning machine model (ELM). The best parameters for each model were found using a grid search with five-fold cross-validation. We

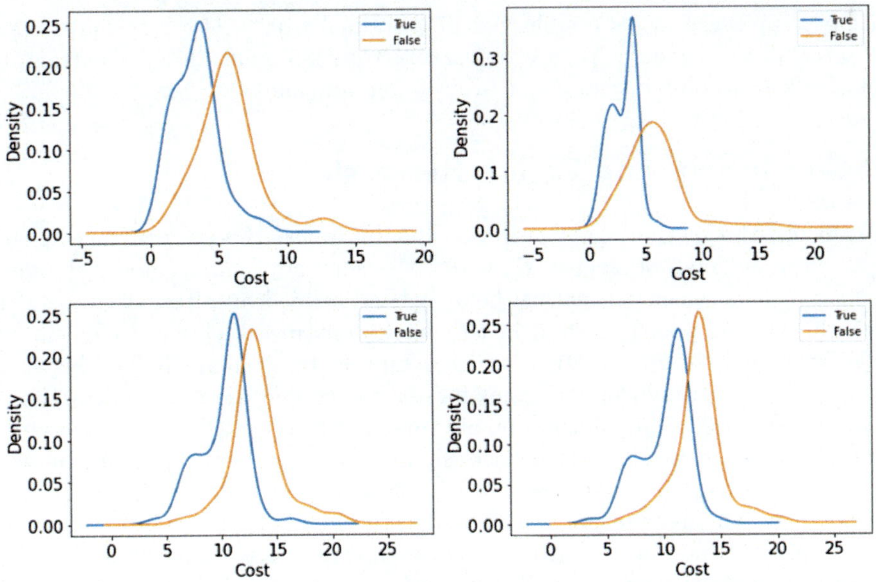

Fig. 3. Comparison between the distribution of features for one specific user in both axes (top in x and bottom in y). The threshold for the left figures is 0.3 and right ones is 0.6.

have used the scikit-learn python library [14] and its compatible XGBoost [5] and scikit-ELM [2] libraries.

Previous research proposed a system based on the DTW algorithm as a method for verifying signatures. Although this method provides a tool to compare time-dependent sequences and therefore in different scales, the method was not "enough" to achieve a "good" performance measure [1]. However, the DTW method can generate another information besides works just as an algorithm of comparison Table 1.

Table 1. Accuracy on the test set for different metrics with different machine learning models for the data including all thresholds in both x and y directions.

Metric	Random forest	XGBoost	ELM
Euclidean	0.82	0.82	0.81
GAK	0.78	0.77	0.72
Kullback-Leibler	0.75	0.73	0.73
Manhattan	0.82	0.82	0.81
Squared	0.79	0.79	0.77

One crucial difference between offline and online systems for signature verification is that the latter can extract additional dynamical features. The simulation of pressure as a possible feature obtained from static signatures opens the possibility to create feature extractors that simulate features extractable exclusively from online systems or bring forth new ones.

7 Conclusions and Future Research

To sum up, the methodology proposed has improved the performance regarding the revised methodologies. However, the accuracy found could not offer a trustworthy system in a practical field; it has outlined an alternative in future research. The proposed method in this work combines a classical element like DTW with five different metrics and machine learning algorithms. This strategy provides a straightforward tool for data augmentation and extraction of prominent features. The information obtained allowed identification patterns, for instance, the cost distribution per threshold and component difference between genuine and forgeries signatures.

Moreover, the distributions of costs do not show relevant differences among users. This was not an essential criterion during the training stage, i.e., the proposed models are incapable of recognizing among users. They were designed to learn how the genuine and the forgeries signatures look, regardless of the user. All studied models present a similar performance on the different metrics, however Manhattan and Euclidean metrics outperforms GAK, Kullback-Leibler and Squared ones.

Despite exhibiting good performance, the obtained results offer an opportunity to improve the measures, producing more data using data augmentation techniques with the same algorithms and even more applying deep learning techniques.

Acknowledgments. The authors wish to acknowledge CSC – IT Center for Science, Finland, for computational resources.

References

1. Acosta-Velásquez, R.D.: Verificación de firmas manuscritas. Master's thesis, Departamento de Matemáticas, Universidad Nacional de Colombia (2013)
2. Akusok, A., Espinosa Leal, L., Björk, K.M., Lendasse, A.: Scikit-elm: an extreme learning machine toolbox for dynamic and scalable learning. In: International Conference on Extreme Learning Machine, pp. 69–78. Springer (2019)
3. Akusok, A., Espinosa Leal, L., Björk, K.M., Lendasse, A., Hu, R.: Handwriting features based detection of fake signatures. In: The 14th PErvasive Technologies Related to Assistive Environments Conference, pp. 86–89 (2021)
4. Bhattacharyya, D., Ranjan, R., Alisherov, F., Choi, M., et al.: Biometric authentication: a review. Int. J. u- e-Serv. Sci. Technol. **2**(3), 13–28 (2009)
5. Chen, T., Guestrin, C.: XGBoost: A scalable tree boosting system. In: Proceedings of the 22nd ACM SIGKDD International Conference on Knowledge Discovery and Data Mining, pp. 785–794. KDD '16, ACM, New York, NY, USA (2016). https://doi.org/10.1145/2939672.2939785, http://doi.acm.org/10.1145/2939672.2939785

6. Cuturi, M.: Fast global alignment kernels. In: Proceedings of the 28th international conference on machine learning (ICML-11), pp. 929–936 (2011)
7. Espinosa-Leal, L., Akusok, A., Lendasse, A., Björk, K.M.: Extreme learning machines for signature verification. In: International Conference on Extreme Learning Machine, pp. 31–40. Springer (2019)
8. Gonzalez, R.C., Woods, R.E., Masters, B.R.: Digital image processing third edition. Pearson International Edition (2008)
9. Hafemann, L.G., Sabourin, R., Oliveira, L.S.: Offline handwritten signature verification - literature review. CoRR abs/1507.07909 (2015). http://arxiv.org/abs/1507.07909
10. Hafemann, L.G., Sabourin, R., Oliveira, L.S.: Learning features for offline handwritten signature verification using deep convolutional neural networks. Pattern Recogn. **70**, 163–176 (2017)
11. Hafemann, L.G., Sabourin, R., Oliveira, L.S.: Offline handwritten signature verification-literature review. In: 2017 Seventh International Conference on Image Processing Theory, Tools and Applications (IPTA), pp. 1–8. IEEE (2017)
12. Ortega-Garcia, J., Fierrez-Aguilar, J., Simon, D., Gonzalez, J., Faundez-Zanuy, M., Espinosa, V., Satue, A., Hernaez, I., Igarza, J.J., Vivaracho, C., et al.: Mcyt baseline corpus: a bimodal biometric database. IEE Proce.-Vis., Image Sig. Process. **150**(6), 395–401 (2003)
13. Padmajadevi, G., Aprameya, K.S.: A review of handwritten signature verification systems and methodologies. In: 2016 International Conference on Electrical, Electronics, and Optimization Techniques (ICEEOT), pp. 3896–3901 (2016)
14. Pedregosa, F., Varoquaux, G., Gramfort, A., Michel, V., Thirion, B., Grisel, O., Blondel, M., Prettenhofer, P., Weiss, R., Dubourg, V., Vanderplas, J., Passos, A., Cournapeau, D., Brucher, M., Perrot, M., Duchesnay, E.: Scikit-learn: Machine learning in python. J. Mach. Learn. Res. **12**, 2825–2830 (2011)
15. Vargas, J.F., Ferrer, M.A., Travieso, C.M., Alonso, J.B.: Off-line signature verification based on high pressure polar distribution. In: Procedeeings of the 11th International Conference on Frontiers in Handwriting Recognition ICFHR, pp. 373–378, (2008)
16. Velichko, V., Zagoruyko, N.: Automatic recognition of 200 words. International J. Man-Mach. Studies **2**(3), 223–234 (1970). https://doi.org/10.1016/S0020-7373(70)80008-6, http://www.sciencedirect.com/science/article/pii/S0020737370800086

Extreme Learning Machine-Based Operational State Recognition: A Feasibility Study with Mechanical Vibration Data

Jukka Junttila[1,2(✉)], Ville S. Lämsä[1], and Leonardo Espinosa-Leal[2]

[1] VTT Technical Research Centre of Finland, P.O. Box 1000, 02044 Espoo, Finland
ville.s.lamsa@vtt.fi

[2] Department of Business Management and Analytics, Arcada University of Applied Sciences, Jan-Magnus Janssons plats 1, 00560 Helsinki, Finland
jukka.junttila@vtt.fi, leonardo.espinosaleal@arcada.fi

Abstract. The benefits of digital twins and accurate near real-time on-site condition monitoring of heavy machinery or load-bearing structures are undeniable. Both demand computationally light and accurate models based on continuously measured data. Extreme Learning Machine (ELM) algorithm provides the means for building accurate and fast predicting classification models. Therefore, the feasibility of the ELM algorithm for building models for near real-time operational state recognition of a rotating machine was studied. Three different models, called one, two, and six cycles, built using the ELM algorithm were compared with corresponding models trained using Support Vector Machine (SVM) and linear regression (LR) algorithms based on their accuracy and prediction times. The comparisons show that the SVM algorithm produces the best accuracy, but with the cost of high prediction times. The LR models have the lowest prediction time. In contrast, the ELM model for the two cycles presents better performance than the corresponding LR and SVM models when the combination of accuracy and the prediction time is considered. The great benefit of the ELM method comes from its mathematical properties: new data can be added to the ELM model without the need to retrain the whole model, and the model is competent to take strong nonlinearities into account. Thus, the possibilities of the ELM algorithm to act as a novelty detector in operational state recognition shall be investigated.

Keywords: State recognition · Vibration data · Extreme learning machine

1 Introduction

By definition, a digital twin (DT) is a virtual real-time representation of its real-world physical counterpart [4]. The more accurate the representation, the better.

K.-M. Björk (Ed.): ELM 2021, PALO 16, pp. 114–123, 2023.
https://doi.org/10.1007/978-3-031-21678-7_11

Typically, DTs are based on measured data of the product/process/system/thing itself. The measured data has to be processed and converted to valuable information using some model. Data processing and, preferably, the conversion to helpful information should be done locally, if possible, to avoid excess data transferring. Therefore, the need for accurate and fast-performing data processing and models is evident.

1.1 Related Works

Improvements in sensors technology have shown high sensitivity, so they are used extensively to monitor operational states in heavy machinery. This method has become necessary for enhancing reliability in the industrial field, mainly for its failure diagnosis and malfunctioning prediction capabilities, among other properties. Despite advances in miniaturization and computational processing, finding the appropriate machine learning models for real applications can be challenging [12]. Linear models are light and robust and are well known to work in many real applications; however, several applications include analysis of nonlinear data, particularly vibration data, where their predictive performance could be compromised in the long term. Support Vector Machines (SVM) are a more suitable method for nonlinear data modeling; unfortunately, this method scales poorly with the size of the data. More recently, an extreme learning machine (ELM) method has become an alternative for modeling big data with an excellent performance between accuracy and scaling [5,6]. Previous research has shown that ELM is superior to SVM when large datasets are modeled [8].

2 Research Methodology

The models built in this study are based on mechanical vibration data measured from a gas engine energy-generating set (genset). The data was measured using triaxial accelerometers at four different locations on a Wärtsilä 20V31SG genset during a set of performance test runs between 19th and 30th of September 2019 in Vaasa, Finland. A Wärtsilä 20V31SG genset is presented in Fig. 1. During the test runs, the genset was tested against different fault situations; the genset was driven at different levels of constant power output, during which sudden changes in the loading of the genset were introduced. Therefore, the measured data includes both normal and abnormal operations of the genset.

This study focuses on classifying and predicting the different states of the regular operation of the genset. Normal operation of the genset is denoted by the constant electrical power output of the genset over a period of time of at least some minutes in length. The different states of normal operation, in this case, are defined by the percentage of the constant power output level of the rated power of the genset. The rated power of the studied genset is approximately 12 MW.

As a result of a previous study [7], it was found out that the power output level of the genset during regular operation can be predicted very accurately from the three signals measured using only one triaxial accelerometer and that the measurement point location affects the classification accuracy. The highest classification accuracy, 99.9 %, was achieved with an SVM algorithm model trained. Three different classification models trained using the SVM algorithm from a previous study were compared with corresponding classification models trained with the ELM and the linear regression (LR) algorithms. In this study, we compared based on their accuracy and prediction time (i.e., the time it takes to make a prediction, not including a training phase). The training and testing of the classification models compared in this study were done using the same data.

Fig. 1. Wärtsilä 20V31SG genset. Image courtesy of Wärtsilä. ©2021 W Wärtsilä corporation.

2.1 Dataset and Features

In the previous study [7], the normal operation data was sampled according to the constant power output level of the genset into the following six different subsets: 0, 50, 75, 90, 95, and 100% of the rated power of the genset. In practice, time intervals of constant operation at the different power levels were identified by visual inspection of the power output graphs produced by the plant automation system software. The measured mechanical vibration signals corresponding to the identified time intervals were stored and grouped by the power output level. Five different algorithms were used to extract features from the stored signals, and their sensitivity to the changes in the power output level of the genset was studied. Finally, a combination of six different features extracted using two different algorithms from the three signals measured at a single measurement point was found to be very effective for solving the classification task.

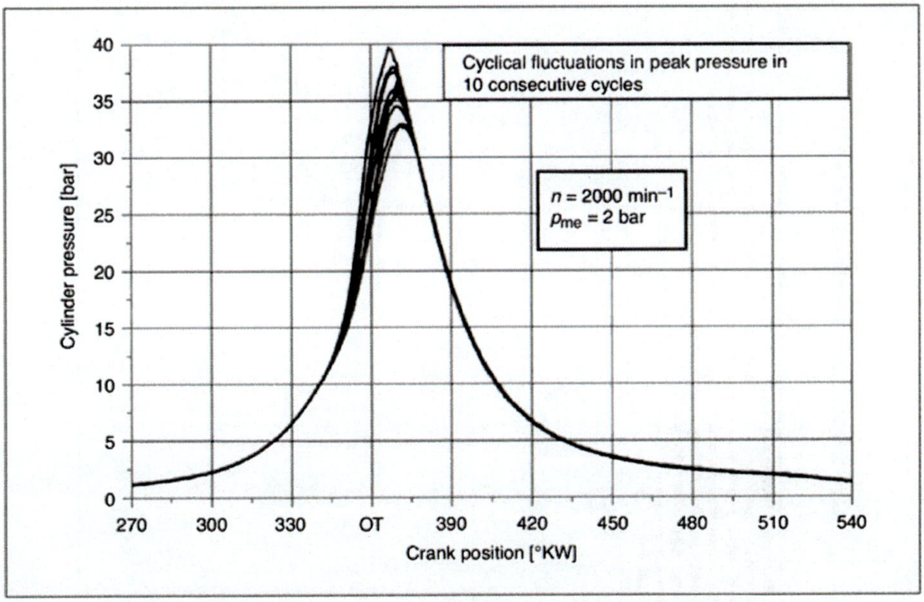

Fig. 2. Cyclic fluctuation in cylinder pressure of a spark ignited engine (adapted from Van Basshuysen, 2016, p. 625) [13].

The algorithms chosen for extracting the features were based on two principal characteristics of an internal combustion engine (ICE) operation at a constant rotation speed. First, the operation of an ICE at constant speed is virtually periodic over engine cycles, although significant fluctuations between cycles may occur. This is especially typical for gas engines, as can be seen from Fig. 2. Therefore, suitable signal processing methods, Fast Fourier transform (FFT) [9] in this particular case, were assumed to give sensitive features. Second, a running ICE generates gas and inertia forces, which are the principal cause of the mechanical vibrations of the genset. At constant rotation speed, the inertia forces of an ICE are also constant, and it is only the gas forces increase as the load increases.

The inertia forces cause an inertial torque on the crankshaft, for which exists an accurate approximation presented in Eq. 1.

$$M = \frac{1}{2} m_{rec} \omega^2 r^2 \left(\frac{r}{2l} \sin \omega t + \sin 2\omega t + \frac{3r}{2l} \sin 3\omega t \right), \tag{1}$$

where m_{rec} is the reciprocating mass, ω is the angular rotation speed, r is the crank radius, and l is the length of the connecting rod [2].

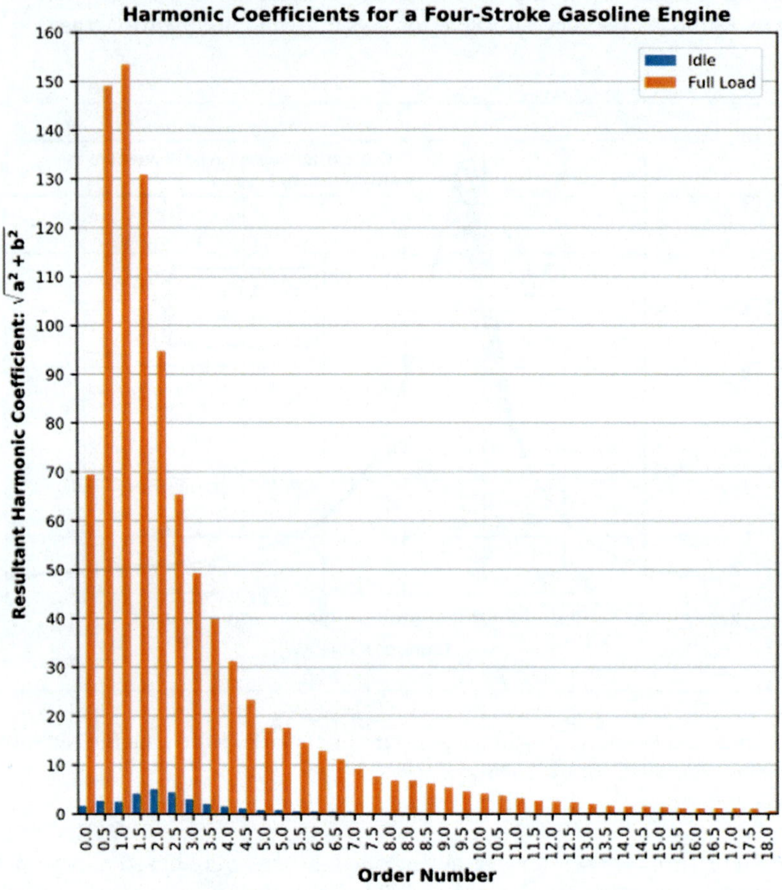

Fig. 3. Resultant harmonic coefficients for the data used in this work [11]. Coefficients **a** and **b** are the respective Fourier coefficients for the cos and sin terms in the Fourier series.

There is no analytical solution for solving the gas forces and the torque caused by them. However, they can be described as a Fourier series by considering the gas forces as periodic over engine cycles. The fundamental frequency of the Fourier decomposition of the gas forces is based on the engine cycle. As the ICE of the studied genset is a four-stroke engine, the base frequency is half of the rotational frequency of the engine, since the four-stroke process takes two revolutions of the crankshaft. Therefore, the so-called order numbers for a four-stroke engine are integer multiples of 0.5: order number 0.5 corresponds to the engine cycle frequency, order number 1.0 corresponds to the rotating frequency, and so on. In his seminal work, *Porter* calculated the harmonic coefficients, **a** and **b**, of the Fourier decomposition of the engine torque due to gas pressure affecting a single crank until the 18th order for different types of engines at different levels of engine load [11]. The resultant harmonic coefficients for a four-stroke gasoline engine at idle (i.e., an engine is running without load) and at full load (i.e., an engine is outputting its maximum power at a specific rotating speed) are presented in Fig. 3.

From Fig. 3 can be seen the difference in the gas forces between idle and full load and that the harmonic coefficients at orders 0.5, 1.0, and 1.5 are the strongest. Equation 1 reveals that the inertia forces have harmonic coefficients only at orders 1.0, 2.0, and 3.0. Therefore, it was assumed and proven in the previous study that the amplitude of the mechanical vibration at order 1.5 would be significantly more sensitive to the changes in the power output level than the amplitude of the mechanical vibration at order 1.0 [7]. It was also found out that the sum of the squared signal values over an engine cycle or over a number of engine cycles is marginally more sensitive to the changes in the power output level of the genset than the sum of absolute signal values over a signal of the same length if it is combined with the mechanical vibration amplitude of the signal at order 1.5 [7].

As shown in Fig. 2 significant fluctuations in the cylinder pressure between consecutive cycles are common in spark-ignited engines. It was assumed and proven in the previous study that extracting the feature values from a longer segment of a signal would even the effect of the fluctuations between cycles to the mechanical vibrations, and hence be more sensitive to the changes in the power output level of the genset. However, the longer the segment, the further away from real-time the prediction becomes. Therefore, the use of segments as short as possible is preferred. It was shown in the previous study that using features extracted from one engine cycle-long segments of the measured signals for predicting the power output of the genset is not as accurate as using features extracted from longer segments [7]. The prediction accuracy increased notably by only doubling the segment length, and the most accurate models were built using a segment length of 6 engine cycles [7]. Therefore, in this study, features extracted from segments of signals of lengths one, two, and six engine cycles were used to build three separate classification models with each of the three classifier algorithms. For each of the three different segment lengths, the dataset used in this study contains 300,000 samples per feature and per class.

2.2 Calculations

Machine learning calculations were performed using the scikit-learn python library [10] and a new compatible library called scikit-ELM [1]. We compared the SVM, the LR, and the ELM algorithms in accuracy and performance. The training of the classifier models was done using stratified five-fold cross-validation. The ELM parameters were optimized by a randomized grid search of 1000 steps. For the other two methods, the parameters were optimized manually [7]. 100,000 samples per class were used for training the final classifier models to be compared, and the remaining 200,000 samples per class were used for testing. The number of samples used for training the classifier models was known to be adequate [7].

All calculations were run in an HPC facility with two nodes with an Intel Xeon Gold 6246 Processor of 24 cores each and 1.5 Tb of memory RAM at the Technical Research Centre of Finland (VTT). The Python 3.9 version was used for all calculations.

3 Results and Discussion

In Table 1 the obtained accuracy and prediction time, denoted as PT, for the three models—one, two, and six cycles—are considered, including the optimal parameters found upon random optimization. The prediction times presented in Table 1 are for the whole test set, that is, 1.2 Million samples, and the unit is seconds.

Table 1. Accuracy on the test set for the three different machine learning algorithms: the extreme learning machine (ELM), the support vector machine (SVM), and the logistic regression (LR). The best parameters obtained upon random grid search and the prediction times (PT) are also included.

Algorithm	Cycles	Best parameters	Accuracy (%)	PT (s)
ELM	1	N = 178, $\alpha = 0.00017$, density = 0.318, original features = False, f = tanh	92.6	4.99
	2	N = 788, $\alpha = 1.1e-5$, density = 0.258, original features = True, f = tanh	98.8	20.37
	6	N = 1648, $\alpha = 1.1e-5$, density = 0.104 original features = False, f = tanh	99.6	32.72
SVM	1	C = 0.1, γ = 1e10	96.6	4428.04
	2	C = 10.0, γ = 10	99.4	1583.11
	6	C = 10.0, γ = 10	99.9	283.02
LR	1	max. iterations = 1e6	95.1	0.14
	2		98.7	0.14
	6		99.7	0.14

The ELM optimized parameters show that the number of neurons is higher for the six cycles than for the one and two cycles. The fact that the *original parameters* variable is False upon optimization for the obtained ELM models for the one cycle and the six cycles confirms that the data is not linear. Confusion matrices for all trained models are plotted in Fig. 4. At the top of the figure, the ELM results are presented; at the middle, the results for the SVM, and at the bottom, the results for the LR.

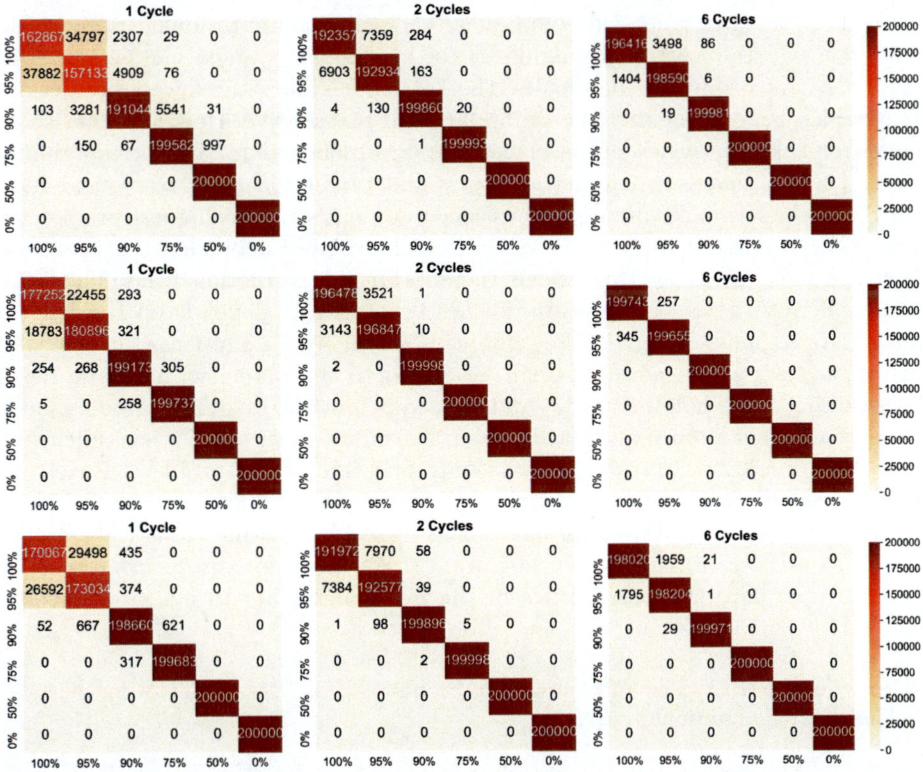

Fig. 4. Confusion matrix for the different machine learning models trained on data obtained at the one, two, and six cycles cycles. *Top:* the extreme learning machine (ELM), *middle:* the support vector machine (SVM), and *bottom:* the logistic regression (LR).

From the confusion matrices, it is easy to see that for the one cycle; the model has difficulties separating the output powers of 100, 95, and 90%. This behavior is similar for all algorithms. Anyhow, the models for the six cycles present a better behavior with the output powers of more than 90%.

4 Conclusion and Future Research

This work showed that it is possible to recognize the different states of the normal operation of the genset via vibrational data measured using an accelerometer. The models showed that when using data based on one engine cycle long segments of signals, it is hard to correctly classify the power output levels above 90%, i.e., different power output levels become mixed. However, when the features are extracted from segments of two or six engine cycles of length, all studied machine learning algorithms present an excellent performance. Thus, that represents the importance of developing methods and models, which can take longer segments of a signal and still enable near real-time computational approaches. Only that way, the essential dynamics of the mechanical systems can be modeled accurately. According to the results, the LR models have the lowest prediction time, which is convenient from the application perspective. However, the ELM model for the two cycles presents better performance than the corresponding LR and SVM models when the accuracy and prediction time are considered. The SVM algorithm results in the best accuracy but with the highest prediction times. The prediction time of an SVM model is highly dependent on its accuracy; the more accurate the model, the shorter the prediction time. Thus, its applicability to near real-time approaches dramatically depends on the length of the data segment. The ELM method can benefit from its intrinsic mathematical properties, where new data can be added to the model without the need of retraining the whole model [3] and its competence to take into account also strongly nonlinear relationships between the input features. The classification of normal operation states represents only one side of the problem. Rarely, a complete set of training data from normal and abnormal operation states exists. Thus, the novelty detection approach shall be used for real-life industrial classification, identification, and state recognition tasks. Therefore, in future research, we aim to explore the possibilities of the ELM algorithm to act as a novelty detector.

Acknowledgments. The authors wish to acknowledge CSC – IT Center for Science, Finland, for computational resources.

References

1. Akusok, A., Espinosa Leal, L., Björk, K.M., Lendasse, A.: Scikit-elm: an extreme learning machine toolbox for dynamic and scalable learning. In: International Conference on Extreme Learning Machine, pp. 69–78. Springer (2019)
2. Den Hartog, J.: Mechanical Vibrations. Dover Publications, New York, United States (1985)
3. Espinosa-Leal, L., Akusok, A., Lendasse, A., Björk, K.M.: Extreme learning machines for signature verification. In: International Conference on Extreme Learning Machine, pp. 31–40. Springer (2019)
4. Grieves, M., Vickers, J.: Digital Twin: Mitigating Unpredictable, Undesirable Emergent Behavior in Complex Systems, pp. 85–113. Springer International Publishing (2017)
5. Huang, G., Huang, G.B., Song, S., You, K.: Trends in extreme learning machines: a review. Neural Networks **61**, 32–48 (2015)
6. Huang, G.B., Wang, D.H., Lan, Y.: Extreme learning machines: a survey. Int. J. Mach. Learn. Cybern. **2**(2), 107–122 (2011)
7. Junttila, J.: Operational state recognition of a rotating machine based on measured mechanical vibration data. Master's thesis, Arcada University of Applied Sciences (2021)
8. Liu, X., Gao, C., Li, P.: A comparative analysis of support vector machines and extreme learning machines. Neural Networks **33**, 58–66 (2012)
9. Nussbaumer, H.J.: The fast fourier transform. In: Fast Fourier Transform and Convolution Algorithms, pp. 80–111. Springer (1981)

10. Pedregosa, F., Varoquaux, G., Gramfort, A., Michel, V., Thirion, B., Grisel, O., Blondel, M., Prettenhofer, P., Weiss, R., Dubourg, V., Vanderplas, J., Passos, A., Cournapeau, D., Brucher, M., Perrot, M., Duchesnay, E.: Scikit-learn: machine learning in python. J. Mach. Learn. Res. **12**, 2825–2830 (2011)
11. Porter, F.P.: Harmonic coefficients of engine torque curves. J. Appl. Mech. **10**(1), A33–A48 (1943)
12. Shafique, M., Theocharides, T., Bouganis, C.S., Hanif, M.A., Khalid, F., Hafız, R., Rehman, S.: An overview of next-generation architectures for machine learning: roadmap, opportunities and challenges in the IoT era. In: 2018 Design, Automation & Test in Europe Conference & Exhibition (DATE), pp. 827–832. IEEE (2018)
13. Van Basshuysen, R., Schäfer, F.: Internal Combustion Engine Handbook—Basics, Components, System, and Perspectives, 2nd edn. SAE International (2016)

Unsupervised Handwritten Signature Verification with Extreme Learning Machines

Anton Akusok[1(✉)], Leonardo Espinosa-Leal[1], Amaury Lendasse[2], and Kaj-Mikael Björk[1,3]

[1] Department of Business Management and Analytics, Arcada University of Applied Sciences, Jan-Magnus Janssons plats 1, 00560 Helsinki, Finland
{anton.akusok,leonardo.espinosaleal,kaj-mikael.bjork}@arcada.fi
[2] University of Houston, Houston, TX 77004, USA
alendass@Central.UH.EDU
[3] Hanken School of Economics, Arkadiagatan 22, 00100 Helsinki, Finland

Abstract. Handwritten signature verification has two approaches based on online or offline data collection, both of them being supervised machine learning tasks. This work investigates the feasibility of unsupervised signature verification. It is inspired by a model-based forged signature generation approach, whose inversion could potentially provide an unsupervised solution for the signature verification task. The model inversion is attempted on a massive collection of image patches taken from samples of a large GPDSS10000 artificial signature verification dataset, pre-processed by a general-purpose deep learning network that extracts 1024 meaningful image features. An Extreme Learning Machine (ELM) solves the inversion problem at a very large scale. The paper proposes practical ways of ELM model structure selection on massive datasets and faster solvers. The results show the feasibility of an unsupervised solution for signature verification.

Keywords: Signature verification · Extreme learning machine · Model inversion · Model structure selection

1 Introduction

Handwritten signature verification [13] is a critical problem, as signatures are widely adopted to verify the validity of various documents - even in electronic format with signature capturing devices. There are two families of approaches [11] for signature verification that differ on the data capture method: online and offline. The online approach [17] captures the data time series at the moment a person is writing a signature. This data includes much personal information regarding the handwriting style of a particular person, making online methods generally more precise than offline ones. The obvious drawback is the necessity to

K.-M. Björk (Ed.): ELM 2021, PALO 16, pp. 124–134, 2023.
https://doi.org/10.1007/978-3-031-21678-7_12

collect the data online at the moment of writing. Other drawbacks include problems with compatibility across various capturing devices, e.g., used in different countries or organizations.

Offline signature verification methods [5] use data in image format, like a photo or a scanned picture. They rely on image processing methods, commonly including deep learning networks of various configurations. Offline methods can process any data equally well, including historical data.

These approaches are similar in supervised methods, i.e., relying on a training set of known valid signatures. This training set is a significant obstacle in practical applications as the ground truth may be unknown. Even if that is not the case, ensuring high data quality on a large scale is a very resource-consuming process.

1.1 Unsupervised Signature Verification

Unsupervised signature verification attempts to tell a forged signature from a real one based on the signature features alone. This idea is inspired by the existence of a generative model for genuine and forged signatures [9]. This model generates a handwriting process for creating a genuine signature from a sequence of characters and generates an imitation process that omits some signature features for creating a skilled forgery of that signature. A method that could reverse this generative model would tell a genuine signature from a forged one without the need for ground truth training data samples.

Previous work [4] hints at the possibility of such a solution. This paper further investigates the limits of an achievable performance by involving a much larger artificially generated dataset of signatures using 3,000 people instead of 75 (with 54 signatures each) and scaling up the predictive model. The authors are aware of a single reference to the unsupervised signature verification [18], with unconvincing results due to a small training data size of only 500 samples.

The technical challenge involves a very large-scale data processing application, with the custom model structure selection approach feasible for such a task. The authors consider the Extreme Learning Machine (ELM) model due to its fast solution and high scaling potential [3]. The paper proposes a large data model structure selection tailored for an ELM model and a faster solver for a very large scale ELM implemented with *scikit-elm*[1] python library.

2 Methodology

2.1 GPDSS-10,000 Dataset

GPDSS 10,000 dataset is a collection of artificially generated signatures from 10,000 different random letter patterns denoted as "users." Its original paper [9] proposes an elaborate motor-level writing model of a well-trained genuine signature that transforms any sequence of letters into a realistic signature image

[1] https://github.com/akusok/scikit-elm.

in some style. Under the proposed model, a skilled forgery simulates a cognitive level forging another person's signature by reading the key points of a genuine signature's writing image, dropping some of the key points, and generating a forged signature from the reduced representation. Each pattern, or "user," in the dataset includes 54 signatures: 24 images with different versions of a generated genuine signature and 30 images of generated skilled forgeries.

This work uses local image features that encode handwriting details while ignoring the shape of a whole signature, a representation suitable for this paper's unsupervised signature verification method. Features from signatures are obtained using a general purpose Inception21k neural network [12], with 1024 outputs taken from the second-last layer [14]. Local features are computed over square sliding windows of sizes $32, 64, 128$ or 256 pixels taken with overlaps of 50 or 90% [6,7].

2.2 Model Structure Selection at Large Scale

Model structure selection changes a model to fit best the problem at hand by tuning its hyperparameters. Hyperparameters are model parameters whose optimal value cannot be calculated from data during model training. Instead, different hyperparameter values produce multiple variants of the same model, whose performance is compared on a validation set. Tuning a single hyperparameter is straightforward, but with many different ones, the search space grows exponentially. The model performance depends on a particular combination of these parameters, making model structure selection a non-trivial task.

Extreme Learning Machine has at least six different hyperparameters to tune. It can be brute-forced by testing millions of random combinations and analyzing the statistics at small problems [1]. The Brute-force method gets infeasible at a very large scale when training a single model takes thousands of CPU hours. So this paper considers only the three important hyperparameters: type of hidden neurons, number of hidden neurons, and L^2 regularization of the output layer, and takes a two-stage approach.

The first stage of model structure selection is finding the optimal hidden neuron type, a very task-specific parameter with a strong influence on performance. Independent of the hidden neuron type, the performance of ELM is limited by a combination of hidden layer size and an L^2 regularization of the output layer. The comparison experiment is set up on a dataset version with 50% overlap that produces much fewer local features per image. The hidden layer size varies from 2 to 4,096 with 12 uniform steps on a logarithmic scale, and the L^2 regularization varies between 10^{-5} and 10^5. An ELM model will not reach its maximum possible performance, being limited to 4096 hidden neurons and a smaller dataset, but the experiment scale is enough for a fair comparison between different types of hidden neurons. Runtime remains feasible as ELM computational complexity $O(l^2N)$ depends on the number of training samples N and the square of the number of hidden neurons l. The best type is selected at the first stage and used afterward.

The first stage uses a simplified performance measure as per-feature classification performance (whether this local feature comes from a genuine or from a forged signature). The area under the curve (AUC) serves as a comparison metric. AUC is selected over accuracy because ELM as a native regression method outputs a continuous value rather than a class, and AUC allows for a fair comparison between solutions with different optimal binary classification threshold levels. Because the comparison is run on a per-feature level instead of per-image, the expected performance is low - however, it would serve the comparison purpose well.

The second stage is the final model training. ELM has a multi-stage computational expression [2] that allows running the most time-consuming computations only once at the largest scale. ELM models with a smaller number of hidden neurons are quickly computed from the saved intermediate representation. The same goes for the solution with different values of the L^2 regularization of the output layer. So the second stage approach is to pre-compute an intermediate model representation at a large scale of 32,768 neurons on the large dataset of sliding windows with 90% overlap using the optimal type of hidden neurons determined in the first stage; then to calculate the solution at various values of L^2 regularization and hidden layer sizes up to 32,768 neurons.

The second stage uses Equal Error Rate [9] (EER), a standard performance metric in signature verification tasks. EER is computed on whole images instead of local image features. Image predictions are calculated as the average predicted value over all local features from that image.

A speedup improvement is implemented in the second stage by using an LSQR algorithm [16] for the output layer solution. This is a finite iterative algorithm for solving systems of linear equations in the least square sense. Generally, it has a similar performance to other direct solvers like Cholesky decomposition-based ones. Nevertheless, for ill-defined systems with degenerate solutions, the LSQR algorithm converges to that solution very fast, while the general solvers take the same time regardless of the linear system properties. The same can be said for rank-deficit systems, where the underlying solution has fewer dimensions than the linear system that encodes it. As both cases are expected to be seen in the analysis, e.g., with a very large L^2 regularization parameter, the LSQR algorithm reduces the overall computation time.

2.3 Experimental Evaluation

The final experiment involves multiple data processing steps, with the largest feasible ELM model that learns the inverse of the fake signature generation model, which is summarized in Fig. 1. The dataset includes 3,000 signature patterns, or "users" (they are not real people like in other similar datasets [15] but serve the same role). Each user has 24 genuine signature images and 30 skilled forgeries, for a total of 54 images per user. Signature images are rendered with high quality at a simulated 600 dpi. Local features are sampled from the signature images by square sliding windows of sizes 32, 64, 128, and 256, uniformly with 90% overlap;

empty sliding windows that don't capture any lines are discarded. A general purpose deep learning model computes 1024 real-valued features from each sliding window. We use 3,000 users out of 10,000 total because of the restrictions on inference times and data storage; even 3,000 users take months of GPU inference time and 1TB of storage space for extracted deep learning features.

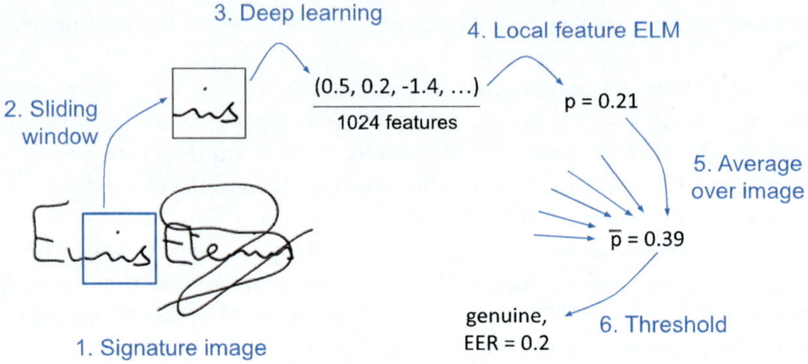

Fig. 1. Experimental workflow with ELM model.

Each local image feature is assigned the binary class of its original signature image: either genuine or forged. Then a large ELM model learns to distinguish local features from genuine signatures from local features from forged signatures. Note that ELM predictions are taken as raw, real value numbers computed by the model; they are not converted back into binary values like the training targets. The expected performance is low due to the task formulation but is significantly above random guessing. The same task is used in model structure selection. An ELM model is trained on 2,000 users. Then the local feature predictions are computed for the remaining 1,000 users, at different values of the L^2 regularization parameter and for different numbers of hidden neurons, as explained in the previous section.

The final prediction for the whole image is an average predicted value of all its local features that is thresholded to convert into a binary prediction. The final metric is Equal Error Rate (EER) [9]. The final model computes binary predictions for 1,000 users * 54 images = 54,000 images, a large number of test samples that allows for a precise method performance evaluation.

3 Results

3.1 Model Structure Selection

Model structure selection is run on a smaller version of the dataset, where the sliding windows for local feature sampling have an overlap of 50%. The dataset still offers 100GB of numerical data for analysis, allowing for a precise comparison between different hyperparameter values.

The optimal type of hidden neurons is the main question of the model structure selection. The comparison is made between Random Projection-based neurons followed by a hyperbolic tangent non-linear function (denoted as *tanh*), and two types of Pairwise function-based neurons using L^1 and cosine distances (denoted as *cityblock* and *cosine*). Only one type of non-linear function is considered in the pairwise projection neurons, as in the authors' experience, the difference in model performance between non-linear functions is rather small. The input space has a large dimensionality at 1024 features that may hide specific feature interactions in a dense random projection involving all 1024 features, so a second option is added for the Random Projection-based neurons that use a sparse random projection with a density of 0.01, effectively ignoring a random subset of 99% input features for each hidden neuron (denoted as *sparse projection*). The Area Under Curve (AUC) of the local feature classification task for these four types of hidden neurons on a smaller dataset is presented in Fig. 2.

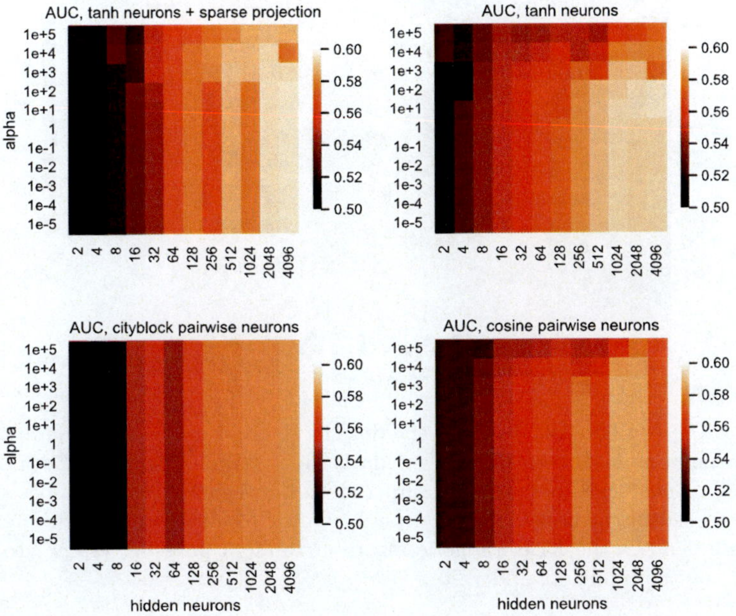

Fig. 2. Area Under Curve (AUC) for 4 different types of hidden neurons on a smaller dataset, for various values of L^2 regularization parameter *alpha* and for different numbers of hidden neurons (limited at 4096 for feasible run times). Color scale denotes model prediction performance, varied from random guessing (*black*) to the best achievable on a per-local feature classification task (*beige*).

The results show that pairwise function based neurons perform worse than random projection based ones. The performance of dense and sparse projections in random projection hidden neurons is similar, with sparse projection having a bit higher AUC and less dependency on the L^2 regularization parameter *alpha*, although at the cost of higher randomness of the model representation and thus model performance. Sparse projection has another benefit of being faster to compute than the dense projection. For those reasons, random projection hidden neurons with sparse projection are selected for the final model.

3.2 Large Scale Experiment

The final experiment is run at the CSC facilities, in parallel on different machines, each handling a solution and evaluation of a model sequence for a fixed L^2 regularization parameter *alpha* but for different numbers of hidden neurons spaced equally on a logarithmic scale from 2 to 32,768. Per-image results are obtained as described in the Methodology section. The average EER over 54,000 validation images for each combination of the hyperparameters is shown in Fig. 3.

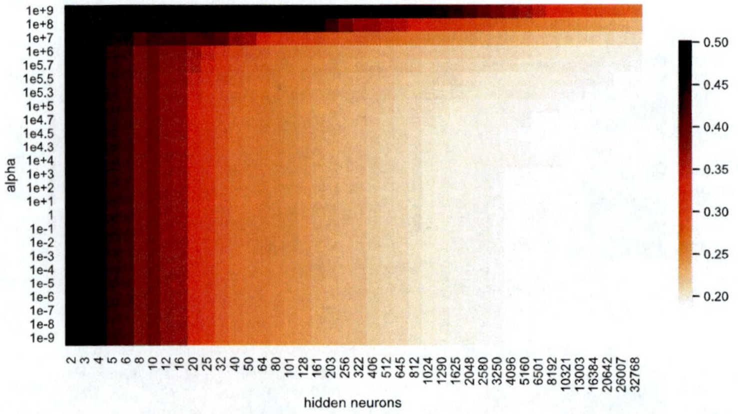

Fig. 3. EER over 1,000 test users for the final large ELM model, for different numbers of hidden neurons and values of L^2 regularization parameter *alpha*. Both scales are logarithmic, *alpha* scale includes extra finer steps between $10^4 \ldots 10^6$ values that highlight the transition to best achievable performance - better seen on the next figure. Color scale denotes the EER range going from random guessing (*black*) to the best achievable by the model at under 20% (*white*).

The performance grows with the number of neurons. Regularization parameter *alpha* works as intended, reducing the model performance at a very large value by applying too much regularization leading to an over-simplified model. Interestingly, unlike in smaller datasets, there is absolutely no "minimum required" regularization on either the number of hidden neurons. At a huge scale of this experiment, the raw number of training samples serves as a sufficient regularization by itself, negating the need for an external regularization. This fact is known theoretically, but here it is demonstrated in practice.

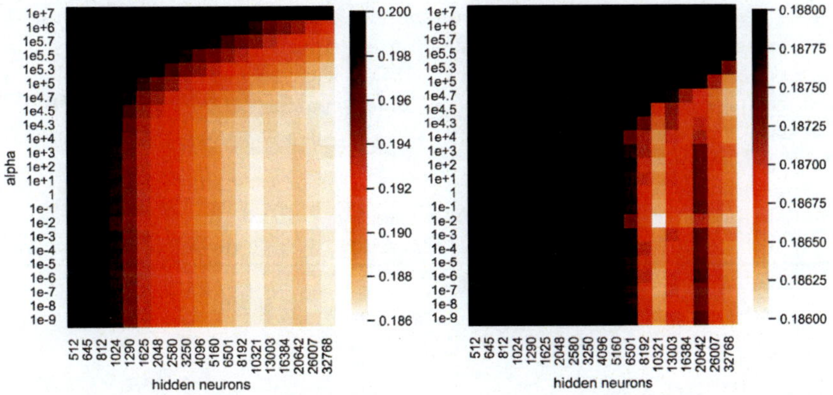

Fig. 4. The zoomed version of Fig. 3 on the model parameter region with the best performance. Color scale denotes EER, axes show number of hidden neurons and the L^2 regularization parameter *alpha* on logarithmic scale. There is an outlier at 10,321 hidden neurons and the *alpha* value of 10^{-2}, but the colors on the figures exaggerate its significance. Model performance is very stable across different hyperparameter values: most EER on the right figure fit in only a 0.1% EER range between [18.725%, 18.625%].

A zoom on the areas with the lowest EER in a lower right corner of the performance chart is shown in Fig. 4. The best model performance stays around 18.7% EER, leaving aside an outlier value on the plots. Reaching the best performance requires 8,000 hidden neurons. However, adding more neurons does not improve the model performance further. This suggests that the experiment has extracted all available information from the dataset in this experimental setup for the ELM model. The deeper zoom on the right of Fig. 4 shows that between 8,192 and 32,768 hidden neurons, EER value fluctuates within $18.7 \pm 0.1\%$. The model performance fluctuation of $\pm 0.1\%$ is minimal (the effect is exaggerated on the plot), so the ELM model approach has reached stable peak performance in the experiment.

The final results are summarized in Table 1. ELM performs worse than the comparison methods, but those methods have access to supervised data with five known genuine signatures per user. ELM is tested on users for whom it has not seen a single labeled signature, whether genuine or forged one. It is surprising to see an unsupervised ELM model performing at the same order of magnitude as alternative supervised methods. These findings suggest that a large part of the information about signature validity is encoded in the way a signature is written, independently of its overall shape, and this information is sufficient to make a judgment about signature validity without access to its known genuine samples - although having access to such samples improves the prediction performance. This was the hypothesis behind this paper, and the experimental results confirmed it.

Table 1. Performance comparison of offline signature verification models for skilled forgeries in terms of EER, trained on 2,000 users from GPDSS 10,000 dataset. Alternative models are hidden Markov model (HMM) with geometrical features [8], and a support vector machine (SMV) with local binary pattern (LBP) features [10]. Comparison data from the GPDSS 10,000 dataset description page at https://gpds.ulpgc.es.

Model	HMM with geometric features	SVM with LBP features	ELM (unsupervised) with deep learning features
EER, %	11.34	16.63	18.7

4 Conclusions

This paper investigates the feasibility of an unsupervised approach to offline signature verification, where no labeled signature samples exist for a given user. The inspiration comes from a generative model [9] that creates both genuine handwritten signatures and their skilled forgeries.

The methodology is based on local image features taken by a sliding window and encoded with a general-purpose deep learning network [12]. Local features are designed to ignore the overall signature shape and extract information about the way a signature is written instead. They form the basis for a user-agnostic signature verification approach.

A large ELM model then classifies local image features as being sampled from a genuine signature or from a skilled forgery. This problem is formulated in a way that does not allow for high classification accuracy, but the performance is significantly higher than random. A proposed way of large-scale model structure selection helps find a model suitable for this task. The final model is trained on a huge dataset with 1TB of local image features data. Two interesting findings highlight that regularization becomes irrelevant at a huge model scale. Finding the model saturation point at which it extracted all relevant information from the dataset and further upscaling does not improve the validation results.

Signatures of test users are classified by taking the mean predicted value of their local image samples and comparing it to a threshold. Test users are fully excluded from the training dataset, and the evaluation considers all signatures related to testing users: both genuine and skilled forgeries.

The results show surprisingly high performance for an unsupervised method that is worse but comparable to the supervised signature verification. These findings suggest that a large portion of signature verification information is encoded in the way a signature is written, independent of its overall shape. That makes unsupervised methods viable signature verification tools in the absence of known genuine signature samples from a particular user.

Acknowledgments. The authors wish to acknowledge CSC – IT Center for Science, Finland, for computational resources.

References

1. Akusok, A., Björk, K.M., Estévez, V.: Randomized model structure selection approach for extreme learning machine applied to acid sulfate soils detection. In: To Appear in International Conference on Extreme Learning Machine. Springer (2021)
2. Akusok, A., Björk, K.M., Miche, Y., Lendasse, A.: High-performance extreme learning machines: a complete toolbox for big data applications. IEEE Access **3**, 1011–1025 (2015)
3. Akusok, A., Espinosa Leal, L., Björk, K.M., Lendasse, A.: Scikit-elm: an extreme learning machine toolbox for dynamic and scalable learning. In: International Conference on Extreme Learning Machine, pp. 69–78. Springer (2019)
4. Akusok, A., Espinosa Leal, L., Björk, K.M., Lendasse, A., Hu, R.: Handwriting features based detection of fake signatures. In: The 14th Pervasive Technologies Related to Assistive Environments Conference, pp. 86–89 (2021)
5. Diaz-Cabrera, M., Morales, A., Ferrer, M.A.: Emerging issues for static handwritten signature biometric. In: Advances in Digital Handwritten Signature Processing. A Human Artefact for e-Society, pp. 111–122 (2014)
6. Espinosa-Leal, L., Akusok, A., Lendasse, A., Björk, K.M.: Extreme learning machines for signature verification. In: International Conference on Extreme Learning Machine, pp. 31–40. Springer (2019)
7. Espinosa-Leal, L., Akusok, A., Lendasse, A., Björk, K.M.: Website classification from webpage renders. In: International Conference on Extreme Learning Machine, pp. 41–50. Springer (2019)
8. Ferrer, M.A., Alonso, J.B., Travieso, C.M.: Offline geometric parameters for automatic signature verification using fixed-point arithmetic. IEEE Trans. Pattern Anal. Mach. Intell. **27**(6), 993–997 (2005)
9. Ferrer, M.A., Diaz, M., Carmona-Duarte, C., Morales, A.: A behavioral handwriting model for static and dynamic signature synthesis. IEEE Trans. Pattern Anal. Mach. Intell. **39**(6), 1041–1053 (2017)
10. Ferrer, M.A., Vargas, J.F., Morales, A., Ordonez, A.: Robustness of offline signature verification based on gray level features. IEEE Trans. Inf. Forensics Secur. **7**(3), 966–977 (2012)
11. Impedovo, D., Pirlo, G.: Automatic signature verification: the state of the art. IEEE Trans. Syst. Man Cybern. Part C (Applications and Reviews) **38**(5), 609–635 (2008)
12. Ioffe, S., Szegedy, C.: Batch normalization: accelerating deep network training by reducing internal covariate shift. arXiv Preprint (2015). arXiv:1502.03167
13. Jain, A.K., Ross, A.A., Nandakumar, K.: Introduction to Biometrics. Springer Science & Business Media (2011)
14. Leal, L.E., Björk, K.M., Lendasse, A., Akusok, A.: A web page classifier library based on random image content analysis using deep learning. In: Proceedings of the 11th Pervasive Technologies Related to Assistive Environments Conference, pp. 13–16 (2018)
15. Ortega-Garcia, J., Fierrez-Aguilar, J., Simon, D., Gonzalez, J., Faundez-Zanuy, M., Espinosa, V., Satue, A., Hernaez, I., Igarza, J.J., Vivaracho, C., et al.: MCYT baseline corpus: a bimodal biometric database. IEE Proc.-Vis. Image Signal Process. **150**(6), 395–401 (2003)
16. Paige, C.C., Saunders, M.A.: LSQR: an algorithm for sparse linear equations and sparse least squares. ACM Trans. Math. Softw. **8**(1), 43–71 (1982)

17. Sae-Bae, N., Memon, N.: Online signature verification on mobile devices. IEEE Trans. Inf. Forensics Secur. **9**(6), 933–947 (2014)
18. Sharma, M., Khanna, K.: Offline signature verification using supervised and unsupervised neural networks. Int. J. Comput. Sci. Mob. Comput. **3**(7), 425–436 (2014)

Check for updates

ELMShip: An Efficient Ship Classifier Using Extreme Learning Machines

Leonardo Espinosa-Leal(✉) and Amin Majd

Department of Business Management and Analytics, Arcada University of Applied Sciences, Jan-Magnus Janssons plats 1, 00560 Helsinki, Finland
{leonardo.espinosaleal,amin.majd}@arcada.fi

Abstract. A new algorithm for ship classification named ELMShip is presented. The proposed methodology uses an extreme learning machine algorithm trained via features extracted via a pre-trained deep neural network. Here, we used an open available labelled dataset of 6252 images and five different classes. The original data was augmented via the extraction of subimages using squared sliding windows of different sizes and overlapped ratios to improve our calculations. Upon fine-tuning, we found that our best model outperforms the previously published results for all classes in all evaluated metrics.

Keywords: Extreme learning machine · Ship classification · Data augmentation · Computer vision

1 Introduction

The primary transport mode for global trade is ocean shipping: around 90% of traded goods are carried over the waves. As such, the oceans provide the principal transport arteries for global trade. However, this comes with opportunities and challenges. As demand for global freight increases, maritime trade volumes are set to triple to 2050. Maritime transport forms part of a whole cluster of economic activities that can create economic value added [14].

Recently, many research and industry organizations have started developing and designing autonomous ships [13]. These new ships will apply state-of-the-art technologies to the marine environment. The advent of autonomous ships that are unmanned or low-manned will reduce the number of people at risk at sea. Even when autonomous navigation does not reduce the number of accidents, safety at sea will improve. In fact, increased safety is one of the primary perceived drivers for autonomous shipping, although this safety increase has not yet been quantified in academic literature [4]. More recently, Wróbel et al. [23] conclude that their analysis suggests that human errors have caused at least 60% of the accidents. An overview of EMSA shows that for 65% of the recorded accidents,

K.-M. Björk (Ed.): ELM 2021, PALO 16, pp. 135–143, 2023.
https://doi.org/10.1007/978-3-031-21678-7_13

the leading contributing cause is an erroneous human action [5]. In addition, the International Maritime Organization (IMO) requires that autonomous ships must be "at least as safe as conventional ships [7].

Small ship detection is essential in autonomous ship technology and plays an essential role in shipping safety. Since traditional object detection techniques based on the shipborne radar are not qualified for the task of near and small ship detection, deep learning-based image recognition methods based on video surveillance systems can be naturally utilized on autonomous vessels to detect near effectively and small ships. However, a limited number of real-world samples of small ships may fail to train a learning method that can accurately detect small ships in most cases.

That autonomous ships will be best suited for ocean navigation, many applications, and studies of autonomous ships focus on large ships recognition and classification in the marine environment. However, despite ocean navigation, inland river navigation is also an important part of water transportation. In the river or near-shore ocean environments, many surface targets are small and compact targets, such as small fishing boats and bamboo rafts. These small targets often move elusively within relatively narrow channels and are easily discriminated against in the complex inland river environment.

Computer vision plays a vital role in most autonomous navigation applications. The significant development of deep learning techniques recently provides great opportunities for proposing an effective autonomous ship-oriented small ship detection method. Deep learning, particularly deep neural networks, achieves increasing performances in object detection tasks [18].

In this paper, we trained a computer vision classification model using an extreme learning machine algorithm. We aim to classify different types of ships from an available open dataset.

The remainder of the paper is organized as follows: In Sect. 2 we present a comprehensive state-of-the-art of the available datasets related with ships, in Sect. 3 the research methodology is presented, in Sect. 4 the results of this work are exposed, including a complete discussion. In Sect. 5, the conclusion of our findings and the future ideas are presented.

2 State-of-the-Art

2.1 Available Data-Sets

There are different public datasets for ship detection. In this subsection, some of these datasets are introduced. The datasets are categorized based on the location of the installed camera for collecting the data. For example, the Airbus ship detection data set contains 100,000 images captured by satellites. This dataset provides maritime scenes of optical aerial images from the visible spectrum. The MASATI dataset contains color images in dynamic marine environments, and it can be used to evaluate ship detection methods. Each image may contain one or multiple targets in different weather and illumination conditions. Furthermore, more datasets are collected from the top by drones, such as the Seagull dataset [19].

The Singapore Maritime Dataset [10] consists of videos recorded during daytime and nighttime and provides ground truth labels for every frame of every video, comprising bounding-boxes and object classes for the corresponding bounding box. The dataset is a multi-class classification. SeaShip dataset [20] is a large-scale dataset of ships that are designed for training and evaluating ship object detection algorithms. The dataset currently consists of 31455 images and covers six common ship types (one carrier, bulk cargo carrier, general cargo ship, container ship, fishing boat, and passenger ship). All the images are from about, 10080 real-world video segments acquired by the monitoring cameras in a deployed coastline video surveillance system.

McShips is a multicategory ship dataset, a challenging and large-scale dataset aimed at ship detection and fine-grained categorization. The dataset includes 14,709 annotated images of ships belonging to 6 classes of warships and seven classes of civilian ships.

In this research, we have used Game of Deep Learning: Ship datasets from the Kaggle.[1] The dataset has 6252 images in train and 2680 images in test data. Five different categories of ships are available in this dataset (Cargo, Military, Carrier, Cruise, and Tankers ships). In this work, we have used the data available for training exclusively. Figure 1 illustrates different categories of ships in the dataset and the distribution of classes.

2.2 Computer Vision

Deep learning has made remarkable achievements in image processing, mainly for object detection. There are different deep learning methods that are developed based on different applications and performances. In this subsection, we will mention some of the most relevant approaches that have been used in the area of ship detection. As a typical one-stage detector, SSD [12] processes images in a single network with high efficiency and accuracy. Faster R-CNN [18] is a two-stage detector, which uses RPN (Region Proposal Network) to produce high-quality region proposals and then detects them with Fast R-CNN [17]. Redmon presents a single neural network named YOLO, which abandons anchor boxes and predicts bounding boxes and class probabilities directly from a complete image in one evaluation [15]. YOLO considers object detection as a regression problem to predict bounding boxes and class probabilities. It can be optimized as end-to-end directly with good detection performance. Fast YOLO can process 155 fps. Compared with other state-of-the-art detection algorithms, YOLO makes more localization errors. YOLOV2 [16] is based on YOLO [15]. YOLOV2 removes the fully connected layers from YOLO and uses anchor boxes to predict bounding boxes. The YOLOV2 model can run with various image sizes, and it is easy to make a trade-off between speed and accuracy. YOLOV2 is faster than

[1] https://www.kaggle.com/arpitjain007/game-of-deep-learning-ship-datasets.

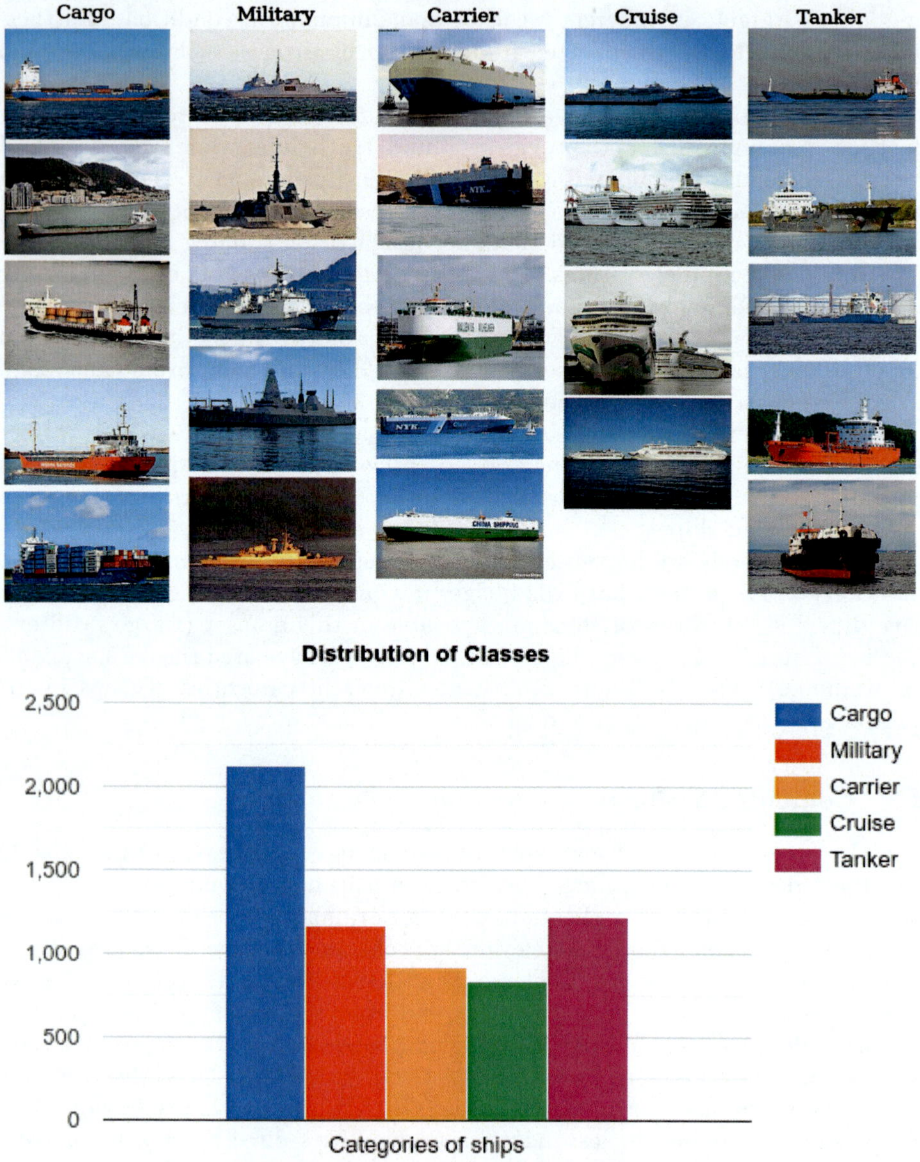

Fig. 1. *Top*: Random selection of images from the Kaggle dataset, presenting the five-different type of ships (classes). *Bottom*: distribution of classes from the Kaggle dataset: Cargo (2120 images), Military (1167 images), Carrier (916 images), Cruise (832 images), and Tanker (1217 images).

YOLO, which can process 200 fps with the Tiny model. Recently, same as other object detection areas, CNNs have been applied for seaborne vessel detection. However, developing more new applications and datasets are required for safe autonomous maritime navigation. For instance, the Singapore Maritime Dataset is used in YOLO for ship detection. In [9], for ship detection, a contextual region-based convolutional neural network with multi-layer fusion has been proposed. It consists of an object detection network with contextual features and a region proposal network (RPN). Their results show that the additional contextual features provide additional information for detection. However, this method cannot detect small objects efficiently. Z. Chen et al. [3] present a novel hybrid deep learning method for small ship detection that combines a modified Generative Adversarial Network (GAN) and a Convolutional Neural Network (CNN)-based detection approach.

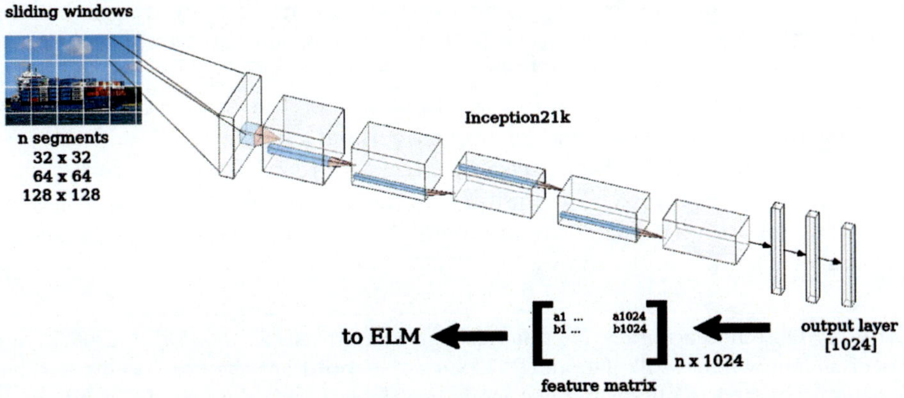

Fig. 2. Diagram of the pipeline for feature extraction. The image is segmented with windows of different sizes and overlaps. Each fragment is passed through a deep network (Inception21k) for feature extraction. Each feature vector is appended to a matrix equal to the class of the original image.

3 Research Methodology

The images' original size in the dataset is less than 210 pixels for the largest side, and the number of classes is five; therefore, we modeled our problem using a multiclass classification approach. The distribution of the classes is relatively balanced (see Fig. 1). However, to find the best model, we augmented the data using sliding windows of different sizes with different overlaps [2,6], Fig. 2 shows a general scheme of the pipeline. We used squared sliding windows of 32, 64, and 128 pixels of side. Moreover, we extracted images including overlaps in all the chosen windows of 0, 10, 25, 50, 75, and 90%. The baseline for our results is the original images.

Different multiclass classification algorithms were trained for all obtained augmented datasets using a state-of-the-art python library for extreme learning machine calculations [1]. The models were feed using features extracted from the images and their sliding segments using a multipurpose trained deep neural network called Inception21k. The optimal parameters (α parameter, density, number of neurons, and the transformation function applied to hidden layer output) for all trained models were obtained using a randomized grid search up to 100 different configurations.

4 Results and Discussion

The results of our experiments are consigned in Table 1. By training our ELM model with the original dataset without performing any augmentation methods, the obtained accuracy was 0.90. Moreover, by using squared sliding windows of 128 pixels of side and overlap of 90%, the obtained accuracy was higher, up to 0.97. The accuracy of the model is much lower when the size of the sliding window decreases. Moreover, a higher overlap tends to improve the scores in general. At the bottom of Table 1 the final size of the dataset upon feature extraction is shown. For the model trained without augmentation and the best model, the values obtained for accuracy, precision, recall, and f1-score per class are summarized in Table 2. The military, carrier, and cruise classes are the best predicted by both models, and despite being the classes with more images, cargo and tanker have the lower scores.

Table 1. Obtained accuracy for different overlaps (0, 10, 25, 50, 75, and 90%) and size of sliding windows (32, 64, and 128 pixels). In **bold** are the best results without augmentation, and the best obtained model. In the last row, the size of the dataset in Gigabytes for each overlap is consigned.

Sliding Window	0%	10%	25%	50%	75%	90%
32	0.45	0.46	0.47	–	–	–
64	0.53	0.57	0.58	0.69	0.76	–
128	0.64	0.66	0.75	0.87	0.94	**0.97**
none	0.89	**0.90**	–	–	–	–
Dataset size	3.1 Gb	3.6 Gb	4.6 Gb	8.4 Gb	28 Gb	100 Gb

For analyzing the performance of our trained models in this paper, a comparison of our results with previous research is presented in Table 3. From our results, it is easy to see that the scores per class obtained via ELM without data augmentation are in line with the best models available in the scientific literature. Our best model outperforms the state-of-the-art for all predicted classes.

5 Conclusion and Future Research

In this paper, an efficient model for ship classification trained using an extreme learning machine was presented. The model used features obtained from images

Table 2. Metric values obtained per each class and overall for a sliding window of 128 pixels and 90% overlap (in parentheses, the results from the dataset without augmentation).

Class	Accuracy	Precision	Recall	f1-score
Cargo	0.97 (0.88)	0.96 (0.85)	0.97 (0.89)	0.96 (0.87)
Military	0.99 (0.96)	0.99 (0.92)	0.99 (0.96)	0.99 (0.94)
Carrier	0.99 (0.95)	1.00 (0.96)	0.99 (0.95)	0.99 (0.95)
Cruise	0.99 (0.92)	0.99 (0.95)	0.99 (0.92)	0.99 (0.94)
Tanker	0.93 (0.77)	0.94 (0.83)	0.93 (0.77)	0.94 (0.80)
overall	0.97 (0.90)	0.98 (0.89)	0.97 (0.89)	0.97 (0.89)

Table 3. Obtained accuracy for different ship classes obtained with ELM and comparison with state-of-the-art algorithms.

Method	Cargo	Military	Cruise	Carrier	Tanker
Hierarchical ship classifier [22]	0.80	-	-	0.93	0.73
Parametric vector estimation + SVM [8]	0.88	-	-	0.80	0.82
Cas-ShipNet [21]	0.88	0.98	0.98	0.99	0.91
Modified ResNet-152 [11]	0.93	0.98	0.98	0.99	0.92
This work (no augmentation)	**0.88**	**0.96**	**0.95**	**0.92**	**0.77**
This work (best model)	**0.97**	**0.99**	**0.99**	**0.99**	**0.93**

available in a public dataset and extracted using a general-purpose deep learning network. We found a straightforward data augmentation using squared sliding windows. Moreover, we found that our results overpass other more elaborated methodologies for predicting the different classes. Future studies using the exposed method in this paper are planned with the much larger and openly available datasets, including a combination of classes. Moreover, we expect to test our results on edge devices for real-time ship detection because the ELM models are light and easy to retrain when adding new data or classes.

Acknowledgments. The authors wish to acknowledge CSC – IT Center for Science, Finland, for computational resources. The authors also thank the *Fonden för teknisk undervisning och forskning* (TUF) for economic support via the *adVanced CompUter vision aLgorithms for the Cognitive modeling of Assistive Autonomous AgeNts* (VULC3AN) project.

References

1. Akusok, A., Espinosa Leal, L., Björk, K.M., Lendasse, A.: Scikit-ELM: an extreme learning machine toolbox for dynamic and scalable learning. In: International Conference on Extreme Learning Machine, pp. 69–78. Springer (2019)
2. Akusok, A., Espinosa Leal, L., Björk, K.M., Lendasse, A., Hu, R.: Handwriting features based detection of fake signatures. In: The 14th Pervasive Technologies Related to Assistive Environments Conference, pp. 86–89 (2021)
3. Chen, Z., Chen, D., Zhang, Y., Cheng, X., Zhang, M., Wu, C.: Deep learning for autonomous ship-oriented small ship detection. Saf. Sci. **130**, 104812 (2020)
4. de Vos, J., Hekkenberg, R.G., Valdez Banda, O.A.: The impact of autonomous ships on safety at sea—a statistical analysis. Reliab. Eng. Syst. Saf. **210**, 107558 (2021). https://doi.org/10.1016/j.ress.2021.107558, https://www.sciencedirect.com/science/article/pii/S0951832021001113
5. EMSA: annual overview of marine casualties and incidents 2019 (2019)
6. Espinosa-Leal, L., Akusok, A., Lendasse, A., Björk, K.M.: Website classification from webpage renders. In: International Conference on Extreme Learning Machine, pp. 41–50. Springer (2019)
7. International Maritime Organization—IMO: autonomous shipping. https://www.imo.org/en/MediaCentre/HotTopics/Pages/Autonomous-shipping.aspx. Accessed 2021-07
8. Jiang, M., Yang, X., Dong, Z., Fang, S., Meng, J.: Ship classification based on superstructure scattering features in SAR images. IEEE Geosci. Remote Sens. Lett. **13**(5), 616–620 (2016)
9. Kang, M., Ji, K., Leng, X., Lin, Z.: Contextual region-based convolutional neural network with multilayer fusion for SAR ship detection. Remote Sens. **9**(8), 860 (2017)
10. Lee, S.J., Roh, M.I., Lee, H.W., Ha, J.S., Woo, I.G.: Image-based ship detection and classification for unmanned surface vehicle using real-time object detection neural networks. In: The 28th International Ocean and Polar Engineering Conference. OnePetro (2018)
11. Leonidas, L.A., Jie, Y.: Ship classification based on improved convolutional neural network architecture for intelligent transport systems. Information **12**(8), 302 (2021)
12. Liu, W., Anguelov, D., Erhan, D., Szegedy, C., Reed, S., Fu, C.Y., Berg, A.C.: SSD: single shot multibox detector. In: European conference on computer vision, pp. 21–37. Springer (2016)
13. NTNU: Autoferry. https://www.ntnu.edu/au-toferry. Accessed 21 Jan 2020
14. Organisation for economic co-operation and development: ocean shipping and shipbuilding. https://www.oecd.org/ocean/topics/ocean-shipping/. Accessed 2019-11
15. Redmon, J., Divvala, S., Girshick, R., Farhadi, A.: You only look once: unified, real-time object detection. In: Proceedings of the IEEE Conference on Computer Vision and Pattern Recognition, pp. 779–788 (2016)
16. Redmon, J., Farhadi, A.: Yolo9000: better, faster, stronger. In: Proceedings of the IEEE Conference on Computer Vision and Pattern Recognition, pp. 7263–7271 (2017)
17. Ren, S., He, K., Girshick, R., Sun, J.: Faster R-CNN: towards real-time object detection with region proposal networks. In: Advances in Neural Information Processing Systems, vol. 28, pp. 91–99 (2015)

18. Ren, S., He, K., Girshick, R., Sun, J.: Faster R-CNN: towards real-time object detection with region proposal networks. IEEE Trans. Pattern Anal. Mach. Intell. **39**(6), 1137–1149 (2016)
19. Ribeiro, R.: Seagull dataset. https://vislab.isr.tecnico.ulisboa.pt/seagull-dataset/. Accessed 2017
20. Shao, Z., Wu, W., Wang, Z., Du, W., Li, C.: Seaships: a large-scale precisely annotated dataset for ship detection. IEEE Trans. Multimedia **20**(10), 2593–2604 (2018)
21. Ucar, F., Korkmaz, D.: A novel ship classification network with cascade deep features for line-of-sight sea data. Mach. Vis. Appl. **32**(3), 1–15 (2021). https://doi.org/10.1007/s00138-021-01198-2
22. Wang, C., Zhang, H., Wu, F., Jiang, S., Zhang, B., Tang, Y.: A novel hierarchical ship classifier for COSMO-SkyMed SAR data. IEEE Geosci. Remote Sens. Lett. **11**(2), 484–488 (2013)
23. Wróbel, K., Montewka, J., Kujala, P.: Towards the assessment of potential impact of unmanned vessels on maritime transportation safety. Reliab. Eng. Syst. Saf. **165**, 155–169 (2017)

An Extreme Learning Machine Model for Venue Presence Detection

Wiqar Khan[1(✉)], Asif Raza[2], Heidi Kuusniemi[3], Mohammed Elmusrati[3], and Leonardo Espinosa-Leal[2]

[1] Cloud and Network Services Nokia, 02610 Espoo, Finland
Wiqar.Khan@Nokia.com

[2] Department of Business Management and Analytics, Arcada University of Applied Sciences, Jan-Magnus Janssons plats 1, 00560 Helsinki, Finland
leonardo.espinosaleal@arcada.fi

[3] University of Vaasa, Vaasa, Finland
{Heidi.Kuusniemi,Mohammed.Elmusrati}@uwasa.fi

Abstract. Value-added services allocation or denial in a particular venue for a given user is of high significance. It will get more prominent as we move to 5G and 6G networks' roll out, as we will get other means to have better aids. In this paper, Extreme Learning Machines (ELM) model performance is compared with Support Vector Machines (SVM), K-Nearest Neighbors (KNN), Logistic Regression (LR), and Random Forest (RF) models for venue presence detection. The input data is collected from the number of UEs (User Equipment) simultaneously placed inside and outside a venue and kept for longer duration. UEs logs essential data such as received signal reference power for serving cells and neighbor candidate cells, along with other network information. Our findings show that ELM model performs above 95% accuracy for a count of zero, one, and two neighbors. The results get better as we consider the collected data from more neighbors' cells in our ELM computation.

Keywords: Venue presence detection · Equipment logs · Extreme learning machines · LTE

1 Introduction

User location in a mobile network has always been of high interest for different use cases. A critical aspect of it has been the value-added services being enabled for the users by the venue owner [11]. Indoor positioning has been more demanding because of the multi-path propagation of communication signals. Most of the venues are considered indoor; therefore, indoor positioning is the subject of interest. Different techniques have been deployed for positioning in indoor facilities. Such techniques range from indoor satellite techniques [13] to mobile and Wi-Fi networks-based techniques. A good survey about selected indoor positioning is presented in [7].

K.-M. Björk (Ed.): ELM 2021, PALO 16, pp. 144–151, 2023.
https://doi.org/10.1007/978-3-031-21678-7_14

As we move towards 5G and 6G networks, we will have more use cases for the in-door positioning, and the interest level in increasing the accuracy will get higher. In 5G and 6G Networks, we will further mean to have better aids. After such advancements, the use cases governed by approaches mentioned in [11] will be more critical than before. The deployment of machine learning algorithms with different conditions and deliberation is the way forward with these in mind.

Real experimental data is collected from 20 User Equipments (UEs) for a prolonged duration, the venue location and setup in which the data was collected is shown in Fig. 1. The 20 UEs are marked as 1 to 20 per the given layout. Zone 1 is the conference hall, i.e., the venue. Zone 2 represent the area outside the venue where we have possible movement in real life. The data used in this study is similar as used in study [12]. The data merging and preparation is also explained there. On the collected data, we have applied Extreme Learning Machine (ELM) for location classification [6,9] and compared the obtained results with other machine learning methods.

2 Analysis

This section will cover the core concepts of RSRP (Reference Signal Received Power), PCI (Physical Cell Identity), serving cell and neighboring cells in LTE (Long-Term Evolution) network [4]. The analysis approaches and break down of the computation is also described here.

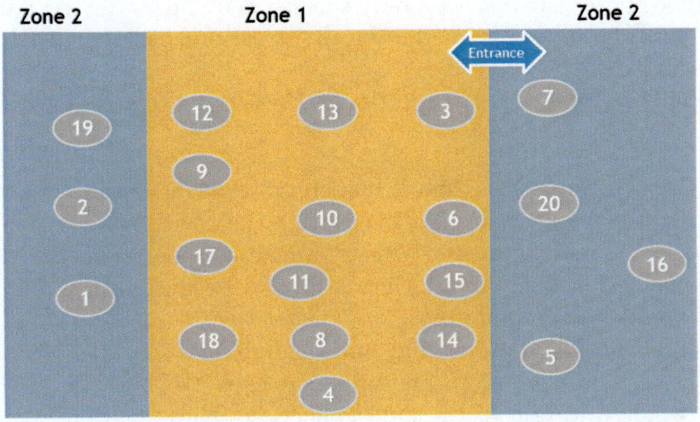

Fig. 1. Scheme of the venue and the user equipment distribution.

2.1 LTE Network Received Data Set

In an LTE network, there is a serving cell where the UE is connected and served by a ‘serving cell’ while there are candidate cells called ‘neighboring cells’ that the UE would connect based on the needs defined per network configurations. In case of need fulfilling the criteria, there could be a handover from the current serving cell to a neighboring cell hence a more suitable neighboring cell will turn into the serving cell while the serving cell will be considered one of the neighboring cells unless there is again a possible handover.

Any cell is identified by its Physical Cell Identify, called PCI. When a UE scans in an LTE network, it records different parameters' values for available PCIs. The main parameter used for cell selection is the RSRP. The RSRP [1] values are obtained for PCIs of the mobile network using the LTE network. There are comprehensive studies available regarding Machine Learning application for Wi-Fi indoor positioning, e.g., [5]. However, we are using RSRPs of the LTE network, and our studied case consists of indoor and outdoor UEs in the same case study as one side of the venue walls to outdoor where UEs were placed for data collection (see Fig. 1).

2.2 Analysis Approaches

In this work, we have considered neighboring cells as well for the learning algorithms, as they provide valuable info about the location where we could get given neighboring cells RSRPs in specific ranges. We use the term neighboring cell depth as the indicator of how many neighboring cells are taken into use and its RSRPs logged by each UE. For our comparison work, we have taken iterations for:

- No neighboring cell (*zero*), i.e., only serving cell and its RSRP data,
- one neighboring cell (*one*) in addition to the serving cell along with respective RSRPs, and
- two neighboring cells (*two*) in addition to the serving cell along with respective RSRPs.

Although there are more than 2 neighbors' cell data used by each UE but in this paper, we considered only 3 scenarios (i.e., without any neighbor, single neighbor and two neighbors). This is aimed to understand the trend as we have more neighbors. Such approach was needed because of the number of ML algorithms being used in our work.

Each UE has its own data logging, and the number of samples collected varies for each UE. It is important to mention that each UE is static at its position during the whole duration of the data collection. The variation of signal levels varies over time based on the mobility around the UEs. The data collection is done in a real-life environment, i.e., when people are moving around, UEs are placed there for a sustained longer duration of around 8–10 h.

3 Extreme Learning Machines

Extreme Learning Machines (ELM) [2] is an emergent machine learning technique, which is used for both classification and regression problems. Extreme learning machine (ELM) has benefit over the Neural Network (NN) on high computation complexity. To overcome the computation complexity of NN, ELM was proposed for the learning of generalized single hidden layer feed forward neural networks, without tuning the hidden layers [10].

In this paper, ELM will be compared against the traditional machine learning approaches (KNN, LR, SVM, RF) to compare the model test time and accuracy. ELM is well suited for a more extensive data set. In the study, [2] author mentioned that "Unlike traditional learning theories and learning algorithms, ELM theories show that hidden neurons need not be tuned in learning and their parameters can be independent of the training data." The ELM architecture benefits from model structure selection and regularization, overfitting, and handing imbalance data.

This study aims to observe the impact of the number of neighbors inclusion in the performance of ELM and other machine learning techniques. Calculations were performed using the state-of-the-art python library for machine learning *scikit-learn* [14], and a new and python compatible extreme learning machine library named *scikit-ELM* [3]. Moreover, with scikit-ELM, we extensively searched the optimal hyperparameters using a randomized gridsearch.

3.1 Parameters Used for ELM

The range of parameters used for the ELM to identify the best parameters on which the model produces the best result are consigned in Table 1. The respective ranges/possible values are given for each parameter.

Table 1. ELM parameters and its ranges.

Parameters	Ranges	Description
alpha	0.00001–1000	Regularization strength
ufunc	*tanh, sigm, relu, lin*	Transformation function of hidden neurons
density	[0.01, 0.99)	Connect hidden neuron to a random input features
Neurons	5–400	Number of hidden layer neurons in ELM model

The optimized parameter *alpha* is the coefficient for the regularization term L^2, *ufunc* the transformation function applied to hidden layer output for adding non-linearity to the algorithm, *density* is the term that controls the density of the random connection weights between input and hidden layers.

4 Results and Discussion

4.1 Comparison of ELM Against ML Algorithms

The detailed comparison result of ELM against conventional ML algorithms is compiled in Table 2. ELM performance is relative. Model test time mentioned in the Table 2 shows that ELM performed better than SVM and KNN. Whereas, results indicate that RF performed better in both time spent on model testing and accuracy where neighbor's count is considered zero.

Table 2. ELM and reference ML algorithms results.

Count of neighbors	ML technique	testing_score	testing_time(s)
	KNN	96.00	34.733
	Logistic Regression	94.50	0.525
zero	Random Forest	96.00	2.076
	Linear SVM	94.50	30.128
	ELM	95.04	10.9
	KNN	97.60	22.142
	Logistic Regression	94.10	0.719
one	Random Forest	96.90	2.458
	Linear SVM	94.50	30.902
	ELM	95.65	11.5
	KNN	99.20	44.898
	Logistic Regression	94.00	0.73
two	Random Forest	98.70	2.705
	Linear SVM	94.40	67.46
	ELM	96.65	10.4

Since the input features increase in the dataset, by adding more neighbors, the time taken by the SVM and LR is significantly increased for model testing where the neighbor count is *one* and *two*, but ELM time spent on model testing remain consistent, which shows that ELM architecture is performing well by spending less time in model testing and produce better results than the fore mentioned algorithms, this also proves that ELM is a better approach where data in imbalance. However, the RF performed better in all the approaches in the data set by both times spent on model testing and accuracy.

Table 3. Neighbors' cells depth and ELM parameters set results. The score is obtained on the test set.

Neighbors	Score (%)	Alpha	Density	Original features	No of neurons	ufunc
zero	95.0420	0.0031	0.0280	TRUE	338	relu
one	95.6505	0.0062	0.6060	TRUE	397	relu
two	96.6519	0.0015	0.5165	FALSE	380	relu

4.2 Impact of Neighbors' Inclusions and ELM Parameters

ELM output values are obtained as we consider further neighbors. The iterations were repeated with the inclusion of neighbors' cells and respective RSRP values. The summary of the results is shown in Table 3. It shows the values of the computed parameters for different neighbors' counts for ELM. As the neighbors' cells depth increases, i.e., as we consider neighbors, the ELM respective accuracy for test results gets better.

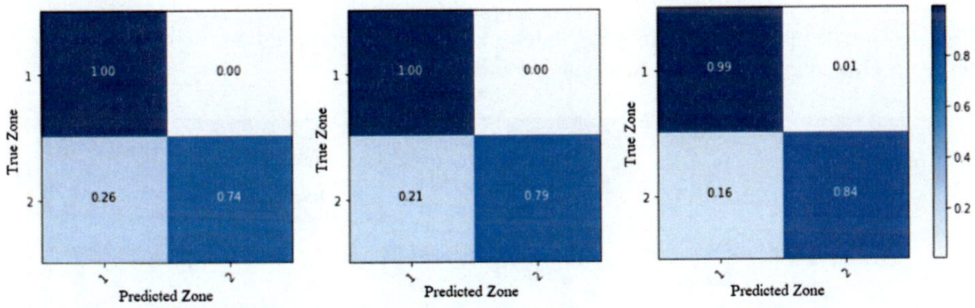

Fig. 2. Confusion matrices for ELM calculations at different count of neighbors: *left: zero*, *middle: one* and, *right: two* cells respectively.

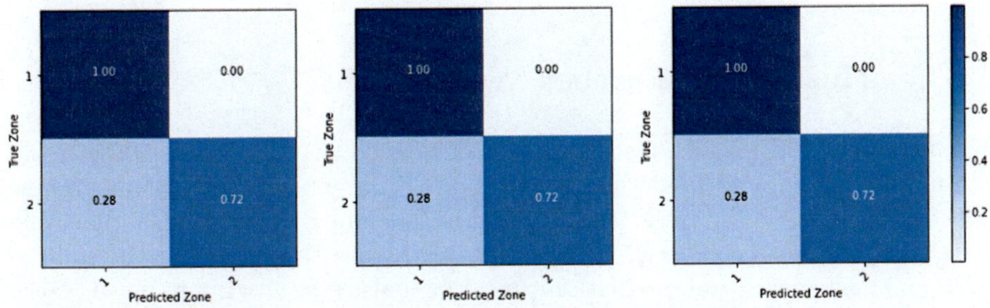

Fig. 3. Confusion matrices for SVM calculations at different count of neighbors: *left: zero*, *middle: one* and, *right: two* cells respectively.

4.3 Confusion Matrices

The obtained confusion matrices for the different machine learning models are depicted in Fig. 2 for ELM, Fig. 3 for SVM, Fig. 4 for RF, and Fig. 5 for LR. In all confusion matrices. The obtained results show that even though the accuracy is almost similar for all trained models, RF, and ELM have fewer false negatives than LR and SVM. The best results are obtained for a count of two neighboring cells.

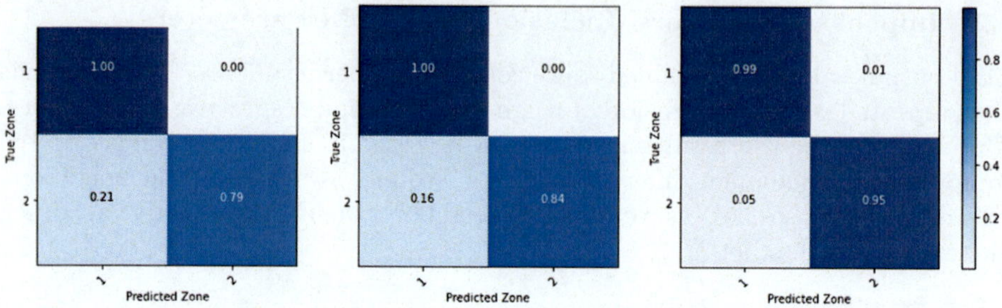

Fig. 4. Confusion matrices for RF calculations at different count of neighbors: *left: zero*, *middle: one*, and *right: two* cells respectively.

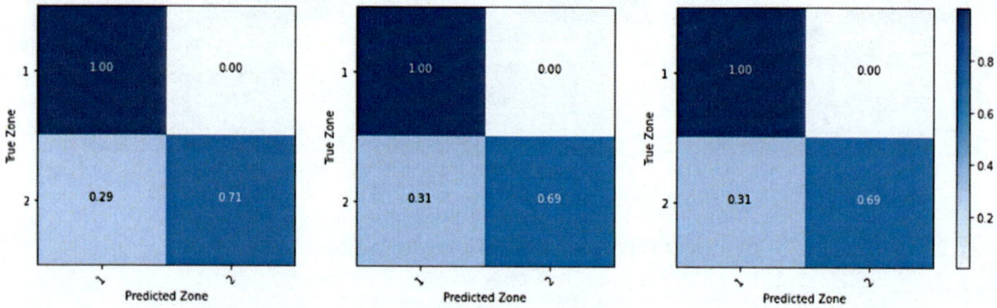

Fig. 5. Confusion matrices for LR calculations at different count of neighbors: *left: zero*, *middle: one*, and *right: two* cells respectively.

5 Conclusion and Further Work

To conclude, ELM is much faster than other techniques used in this study. ELM can handle big data without the need for GPU or larger RAM size. The training and testing time both were fast for this method in comparison to the other tested techniques. In general, ELM is the algorithm that presents the highest ratio among accuracy and prediction time, being also effective in predicting fewer false negatives. Moreover, the advantage of using ELM is that the final models perform excellent with non-linear data and that models can be trained with new data without the need of retraining the whole model each time [3,8].

In this paper, we considered data sets for up to two neighbors. Further research aims to extend it for a higher number of neighbors when the data is available. Moreover, it will be worth deploying the same techniques for 5G networks with beamforming at hand.

Acknowledgments. The authors wish to acknowledge CSC – IT Center for Science, Finland, for computational resources.

References

1. 3GPP: evolved universal terrestrial radio access (E-UTRA); physical layer; measurements. Technical specification (TS) 36.214, 3rd generation partnership project (3GPP) (2018). https://portal.3gpp.org/desktopmodules/Specifications/SpecificationDetails.aspx?specificationId=2440. Version 14.4.0 Release 14
2. Akusok, A., Björk, K.M., Miche, Y., Lendasse, A.: High-performance extreme learning machines: a complete toolbox for big data applications. IEEE Access **3**, 1011–1025 (2015)
3. Akusok, A., Leal, L.E., Björk, K.M., Lendasse, A.: Scikit-elm: an extreme learning machine toolbox for dynamic and scalable learning. In: International Conference on Extreme Learning Machine, pp. 69–78. Springer (2019)
4. Amirijoo, M., Frenger, P., Gunnarsson, F., Kallin, H., Moe, J., Zetterberg, K.: Neighbor cell relation list and physical cell identity self-organization in LTE. In: ICC Workshops-2008 IEEE International Conference on Communications Workshops, pp. 37–41. IEEE (2008)
5. Bozkurt, S., Elibol, G., Gunal, S., Yayan, U.: A comparative study on machine learning algorithms for indoor positioning. In: 2015 International Symposium on Innovations in Intelligent Systems and Applications (INISTA), pp. 1–8. IEEE (2015)
6. Cambria, E., Huang, G.B., Kasun, L.L.C., Zhou, H., Vong, C.M., Lin, J., Yin, J., Cai, Z., Liu, Q., Li, K., et al.: Extreme learning machines [trends & controversies]. IEEE Intell. Syst. **28**(6), 30–59 (2013)
7. Davidson, P., Piché, R.: A survey of selected indoor positioning methods for smartphones. IEEE Commun. Surv. Tutorials **19**(2), 1347–1370 (2016)
8. Espinosa-Leal, L., Akusok, A., Lendasse, A., Björk, K.M.: Extreme learning machines for signature verification. In: International Conference on Extreme Learning Machine, pp. 31–40. Springer (2019)
9. Huang, G.B., Zhu, Q.Y., Siew, C.K.: Extreme learning machine: a new learning scheme of feedforward neural networks. In: 2004 IEEE International Joint Conference on Neural Networks (IEEE Cat. No. 04CH37541), vol. 2, pp. 985–990. IEEE (2004)
10. Huang, G.B., Zhu, Q.Y., Siew, C.K.: Extreme learning machine: a new learning scheme of feedforward neural networks. In: 2004 IEEE International Joint Conference on Neural Networks (IEEE Cat. No.04CH37541). vol. 2, pp. 985–990 (2004). https://doi.org/10.1109/IJCNN.2004.1380068
11. Khan, G.W., Nyman, K.E.O.: Venue owner-controllable per-venue service configuration (2019). US Patent 10 320 973 B2
12. Khan, W., Keskinen, M., Raza, A., Kuusniemi, H., Elmusrati, M.: Using machine learning for in-out decision accuracy for venue owner definable services. In: 2020 International Conference on Communications, Signal Processing, and their Applications (ICCSPA), pp. 1–6. IEEE (2021)
13. Kuusniemi, H., Bhuiyan, M.Z.H., Ström, M., Söderholm, S., Jokitalo, T., Chen, L., Chen, R.: Utilizing pulsed pseudolites and high-sensitivity GNSS for ubiquitous outdoor/indoor satellite navigation. In: 2012 International Conference on Indoor Positioning and Indoor Navigation (IPIN), pp. 1–7. IEEE (2012)
14. Pedregosa, F., Varoquaux, G., Gramfort, A., Michel, V., Thirion, B., Grisel, O., Blondel, M., Prettenhofer, P., Weiss, R., Dubourg, V., et al.: Scikit-learn: machine learning in python. J. Mach. Learn. Res. **12**, 2825–2830 (2011)

Edammo's Extreme AutoML Technology – Benchmarks and Analysis

Brandon Warner[1], Edward Ratner[1](✉), and Amaury Lendasse[2,3]

[1] Edammo Inc, Iowa City, Iowa, USA
ed.ratner@edammoinc.com
[2] Department of ILT, University of Houston, Houston, USA
[3] Arcada University of Applied Sciences, Helsinki, Finland

Abstract. Whether a model is utilizing classical machine learning (ML) algorithms (e.g., Support Vector Machines, KNN, etc.) or neural architectures (e.g., Deep Learning), one must choose a method for selecting optimal hyperparameters. There are both manual and automated methods for hyperparameter optimization. Automated methods are generally referred to as automated machine learning, or AutoML [1]. Several automated selection algorithms have shown similar or improved performance over state-of-the-art methods [2–6]. This breakthrough has led to the development of cloud-based services like Google AutoML [8], which is based on Deep Learning [9] and is widely considered to be the industry leader in AutoML services. We use a fundamentally different type of neural architecture: Extreme Learning Machines (ELMs) [7–13]. ELMs are generally less computationally expensive than Deep Learning. We benchmark our Extreme AutoML technology versus Google's AutoML using four popular classification datasets from the University of California at Irvine's (UCI) repository [27]. We observe significant advantages for Extreme AutoML in accuracy, Jaccard Indices [28], the variance of Jaccard Indices across classes (i.e. class variance) and training times.

Keywords: Machine learning · Deep learning · Extreme learning machines · Classification · Hyperparameter · AutoML · Hyperparameter selection

1 Introduction

There are various ways to go about selecting the structure and parameters of a machine learning model. One popular learner class is Deep Learning [9]. In our approach, we use a fundamentally different learner type called Extreme Learning Machines (ELMs) [10–16]. Whether one is using Deep Learning or ELMs, one must choose a technique for selecting hyperparameters. ELM parameters include the number of neurons, types of neurons, and choice of regularization. One has a few options when choosing a hyperparameter selection technique. The first is to select manually based on the selector's (i.e., Data Scientist) prior experience and expert domain knowledge. Another way is to make selections manually based on trial and error using training and validation sets. A third way is to manually select using techniques that search about the hyperparameter space

K.-M. Björk (Ed.): ELM 2021, PALO 16, pp. 152–163, 2023.
https://doi.org/10.1007/978-3-031-21678-7_15

using one or more numerical optimization techniques. In that case, the user must set the starting point for the search and termination criteria. Each of these techniques require human involvement. There are also automated ways to select optimal hyperparameters. These methods are often referred to as AutoML [1, 5, 6]. Several automated selection algorithms have shown similar or improved performance over state-of-the-art methods [2–4]. Google AutoML [8] is considered by most to be the leader of this field. While Google AutoML employs Deep Learning in their models, Edammo AutoML uses an ensemble of Extreme Learning Machines (ELMs) to operate without the need to set hyperparameters. A number of ELM ensemble techniques have been proposed for varying application domains [17–26]. To best compare these two methods, we chose four popular classification datasets from the UCI repository [27], measuring the methods' performance based on training time, accuracy, Jaccard Indices, class variance, and training times. In Sect. 2, we outline our novel approach to AutoML technology. In Sect. 3, we provide an overview of the datasets and benchmarking methodology. In Sect. 4, we present the results for all datasets and methodologies. In Sect. 5, we briefly restate our results and the future research we plan on conducting.

2 Overview of Novel Approach

In the machine learning domain, a number of metrics are considered when evaluating a model's performance. The significance of each metric and the relative importance will depend on the particular problem at hand. A few of these metrics include accuracy, precision, Jaccard Index variability and training time. Many contemporary machine learning methodologies, such as Deep Learning, require relatively high volumes of data and long training times to make accurate predictions. Extreme AutoML has approached machine learning classification, regression, and time-series prediction in a fundamentally different way. This allows our Extreme AutoML technology to produce fast, accurate models on both small data and large sets. This novel architecture enables the users of our technology to leverage their dynamic data in real time.

Classical and Deep Learning algorithms also require extensive hyperparameter optimization. Extreme AutoML's revolutionary architecture automatically produces highly accurate models without the need to optimize hyperparameters, allowing users to instead focus their efforts on leveraging these state-of-the-art models to their advantage.

Another important consideration to take into account when choosing the best machine learning technology for classification problems is the variability of performance across classes. While a model with overall relatively high accuracy may seem adequate at first, poor performance on one or more of the classes will result in a significant decrease in utility. The metric that best measures performance across classes is the Jaccard Index [28]:

$$J(A, B) = \frac{|A \cap B|}{|A \cup B|} \tag{1}$$

where J is the Jaccard Distance between sets A and B.

Models that produce higher Jaccard Indices for underrepresented classes illustrate greater generalization and utility. In the following sections, we report the results for classification.

3 Datasets and Methodology

The University of California at Irvine's machine learning repository [25] contains thousands of open-source datasets used for classification, regression, time-series, and clustering experiments. To best benchmark Extreme AutoML's technology against state-of-the-art industry leaders, performance comparisons were made on classification models side by side with Google's AutoML service. Metrics measured include accuracy, training time, and Jaccard Indices for each class in the dataset. The first experiment utilized the Human Activity Recognition Using Smartphones dataset [29], which was separated into predefined training testing splits (80% training, 20% testing) and consists of 561 attributes and 10,299 observations. The second dataset, Parkinson's Disease Classification [30], was not split beforehand. Therefore, 5-fold cross-validation was used (five splits of 80% training, 20% testing). The original Parkinson's dataset is 754 variables by 756 observations, but since Google AutoML requires a minimum of 1000 observations to train a model, the dataset was doubled and 0.1% noise was added. The third experiment used the QSAR Oral Toxicity [31] dataset, also using 5-fold cross-validation. However, Google AutoML failed to train on two of the five folds so only the successfully predicted folds were included in the comparison. The fourth dataset, CNAE9[25], failed to train using Google AutoML so the dataset was doubled and 0.1% of noise was added to the training sets. Only one of the five folds successfully trained using Google AutoML so only that fold was included in the comparison.

4 Results

4.1 Human Activity Recognition

Table 1 shows the breakdown of the Human Activity Recognition [29] dataset's attributes and the performance for each learning method. In the first experiment using the Human Activity dataset, Extreme AutoML outperformed Google AutoML by 1.66% accuracy, whilst training the model over 72 times faster. Table 2 shows the various Jaccard Indices for the six classes in the Human Activity Recognition dataset. For each of the classes, Extreme AutoML outperformed Google AutoML in terms of Jaccard Indices, showing greater generalization capabilities (Fig. 1).

Table 1. Results - human activity recognition dataset

Dataset	Method	Training instances	Attributes	Training time (seconds)	Accuracy
Human activity recognition	Extreme AutoML	8239	8239	203	0.9626
Human activity recognition	Google AutoML	8239	8239	14700	0.946

Table 2. Jaccard indices - human activity recognition dataset

Dataset	Method	Jaccard class 1	Jaccard class 2	Jaccard class 3	Jaccard class 4	Jaccard class 5	Jaccard class 6	Class variance
Human activity recognition	Extreme AutoML	0.944	0.9476	0.9718	0.8613	0.875	0.9816	0.00253
Human activity recognition	Google AutoML	0.9301	0.9109	0.9222	0.8585	0.8727	0.9795	0.00188

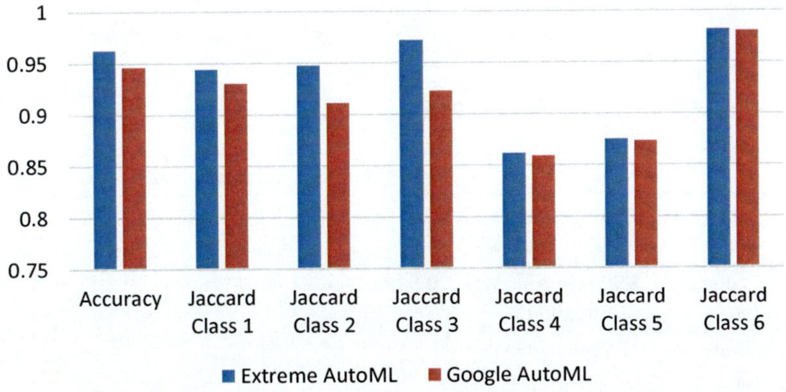

Fig. 1. Classification performance - human activity recognition dataset

4.2 Parkinson's Disease Classification

Edammo's Extreme AutoML has the ability to be optimized for accuracy or speed. In the second experiment using the Parkinson's dataset [30], the Edammo AutoML accurate method outperformed Google AutoML by 1.45% accuracy, whilst training almost 30 times faster. The Edammo AutoML fast method slightly outperformed Google AutoML by 0.2%, but at a significantly faster training speed (291 times faster than Google AutoML). Tables 3 and 4 shows the breakdown of the accuracy and Jaccard Indices for the two classes in the dataset. For each of the classes and methods, Extreme AutoML outperformed Google AutoML in terms of Jaccard Indices, further illustrating greater generalization capabilities (Fig. 2).

4.3 QSAR Oral Toxicity

In the third experiment, using the QSAR Oral Toxicity dataset [31], the Edammo's Extreme AutoML accurate method performed 2.17% better than Google AutoML, whilst training 19 times faster. The Edammo AutoML fast method outperformed Google AutoML by 1.93%, whilst training over 194 times faster. Table 5 shows the breakdown

Table 3. Results – Parkinson's disease classification dataset

Dataset	Method	Training instances	Attributes	Training time (seconds)	Accuracy
Parkinson's disease classification	Extreme AutoML accurate method	1202	752	2050	0.9233
Parkinson's disease classification	Extreme AutoML fast method	1202	752	205	0.9088
Parkinson's disease classification	Google AutoML	1202	752	59820	0.9046

Table 4. Jaccard indices - Parkinson's disease classification dataset

Dataset	Method	Jaccard class 1	Jaccard class 2	Class variance
Parkinson's disease classification	Extreme AutoML accurate method	0.85416	0.86158	2.7528 E-5
Parkinson's disease classification	Extreme AutoML fast method	0.82848	0.8371	3.71522 E-5
Parkinson's disease classification	Google AutoML	0.8217	0.8302	3.6125 E-5

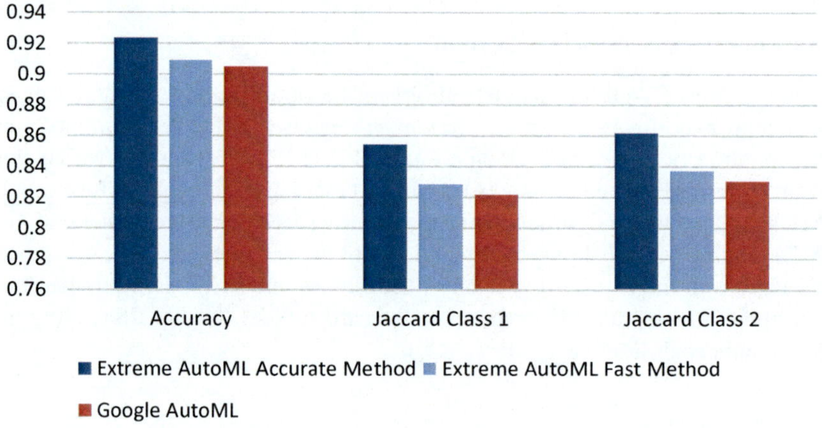

Fig. 2. Classification performance - Parkinson's disease classification dataset

of the Jaccard Indices for each of the classes in the QSAR Oral Toxicity model. For each of the classes and methods, Edammo AutoML outperformed Google AutoML in terms of Jaccard Indices. Most notably, in this unbalanced dataset, the Jaccard Index for the underrepresented class was more than 2x higher for the Edammo AutoML model than the Google AutoML model. This increased generalization capacity is further illustrated in the variance in performance across classes, which is shown in the rightmost column of Table 6 (Fig. 3).

Table 5. Results - QSAR oral toxicity dataset

Dataset	Method	Training Instances	Attributes	Training time (seconds)	Accuracy
QSAR oral toxicity	Extreme AutoML accurate method	4316	700 (Google failed on 999, 900, 800)	1920	0.9394
QSAR oral toxicity	Extreme AutoML fast method	4316	700 (Google failed on 999, 900, 800)	192	0.937
QSAR oral toxicity	Google AutoML	4316	700 (Google failed on 999, 900, 800)	37260	0.9177

Table 6. Jaccard Indices - QSAR Oral Toxicity dataset

Dataset	Method	Jaccard class 1	Jaccard class 2	Class variance
QSAR oral toxicity	Extreme AutoML accurate method	0.3853	0.937	0.1522
QSAR oral toxicity	Extreme AutoML fast method	0.3522	0.934	0.1692
QSAR oral toxicity	Google AutoML	0.1868	0.9161	0.2659

4.4 CNAE9 Business Classification

In the final experiment, using the CNAE9 dataset [27], Edammo AutoML performed 1.4% better than Google AutoML, whilst training 165 times faster. Table 8 shows the breakdown of the Jaccard Indices for each of the classes in the CNAE9 Business Classification model. Edammo AutoML's minimum Jaccard Index was higher than Google AutoML by 0.1, a 20% improvement. The rightmost column in Table 8 shows the expansive gap between the variance across classes for the two services, further illustrating Edammo's superior generalization capabilities (Fig. 4; Table 7).

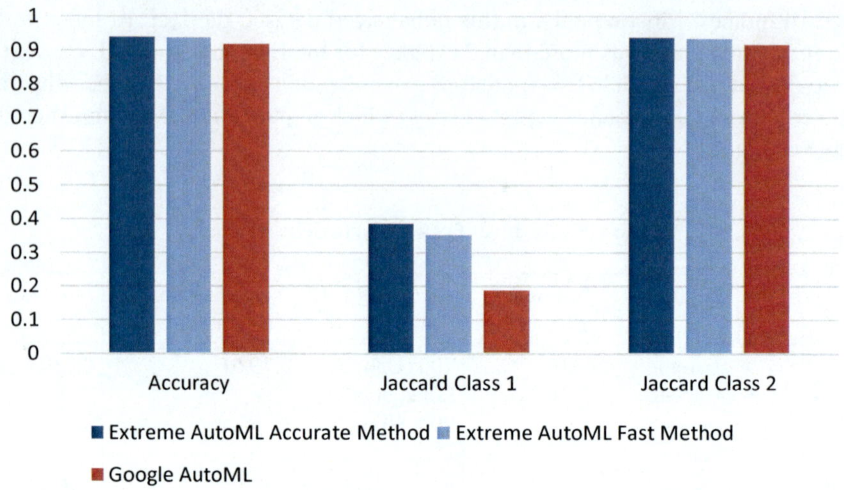

Fig. 3. Classification performance - QSAR oral toxicity dataset

Table 7. Results - CNAE9 business classification dataset

Dataset	Method	Training instances	Attributes	Training time (seconds)	Accuracy
CNAE9 business classification	Extreme AutoML	1728	856	83	0.7721
CNAE9 business classification	Google AutoML	1728	856	13740	0.7581

4.5 Results Summary

In this section, we share the overall classification performance of the two methodologies by comparing relative error rates, training times, and variance (both relative and absolute) of Jaccard Indices across classes for all datasets. We begin by presenting the relative error rates in Fig. 5.

Next, we present a visualization of training times for all datasets and methodologies. Extreme AutoML is not visible for three of the datasets due to the vast difference in training times between the two methodologies (Fig. 6).

Figure 7 illustrates the difference in the absolute variance of the Jaccard Indices for all methodologies and datasets. Extreme AutoML produced a lower variance of Jaccard Indices across classes for each of the datasets when compared to Google AutoML. This further illustrates Extreme AutoML's greater generalization capabilities.

Table 8. Jaccard indices CNAE9 business classification dataset

Dataset	Method	Jaccard class 1	Jaccard class 2	Jaccard class 3	Jaccard class 4	Jaccard class 5	Jaccard class 6	Jaccard class 7	Jaccard class 8	Jaccard class 9	Class variance
CNAE9 business classification	Extreme AutoML	0.7692	0.5454	0.5937	0.5714	0.7619	0.5	0.6667	0.7037	0.6258	0.0095
CNAE9 business classification	Google AutoML	0.7916	0.7407	0.4038	0.5625	1.0	0.4706	0.88	0.7037	0.4	0.046

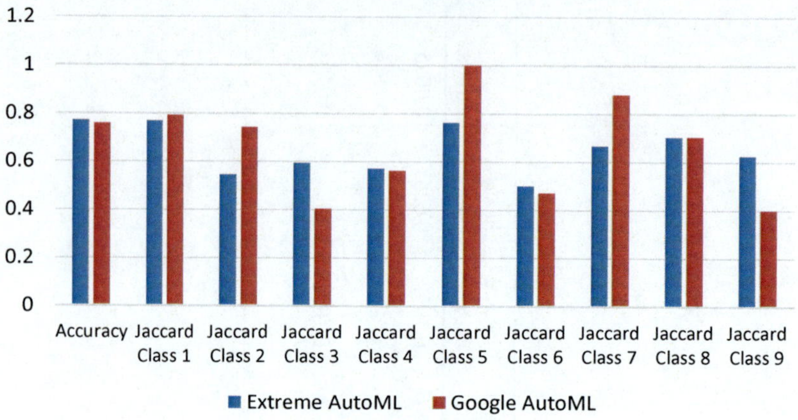

Fig. 4. Classification performance - CNAE9 business classification dataset

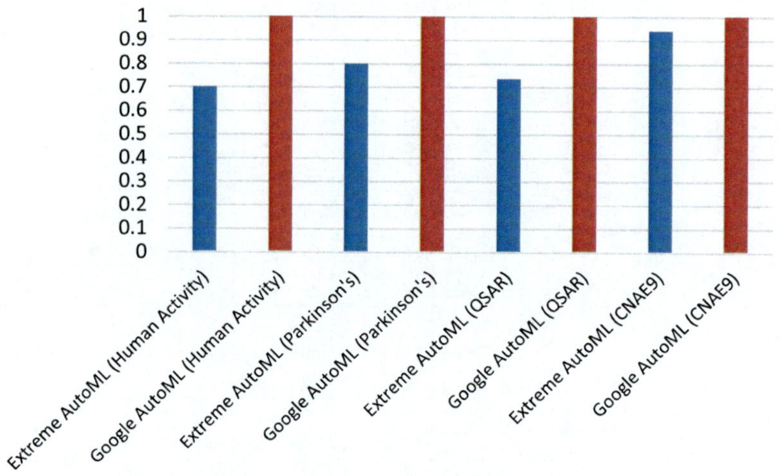

Fig. 5. Relative error rates - all datasets and methodologies

5 Conclusions and Further Work

This paper provides a robust set of benchmarks using ML classification datasets from the UCI repository using two state-of-the-art AutoML methodologies: Google AutoML and Extreme AutoML. We also provide an overview of Edammo AutoML's novel approach, differentiating from the near-ubiquitous Deep Learning paradigm. We show significantly better results on a number of key metrics of interest for the Extreme AutoML approach for classification problems. In future work, we will benchmark Extreme AutoML versus Deep Learning Auto ML approaches for other types of models including generalized regression and time series.

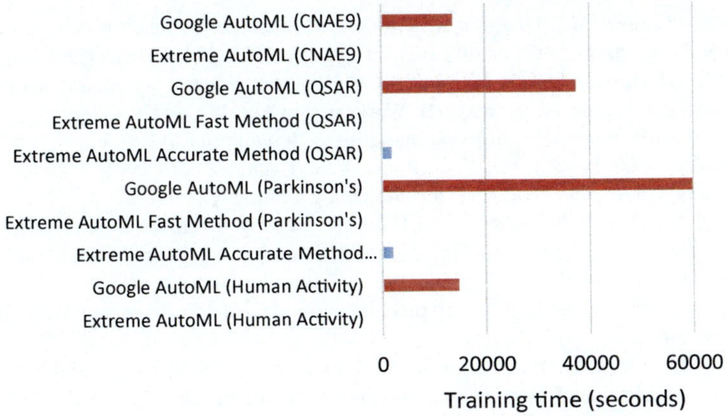

Fig. 6. Training times for all datasets and methodologies

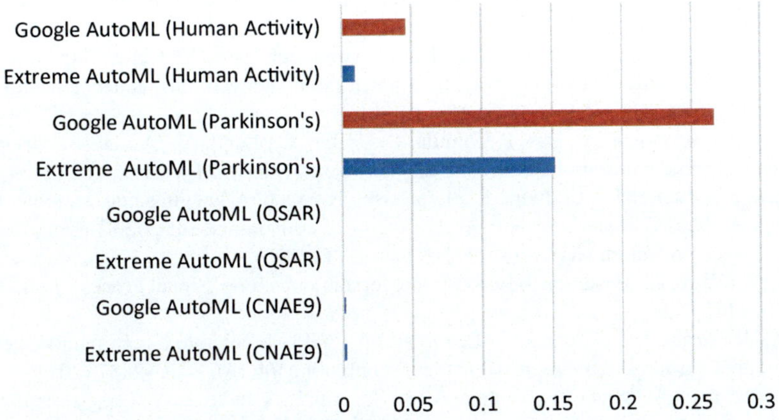

Fig. 7. Absolute variance of Jaccard indices across classes for all datasets

References

1. Yao, Q., Wang, M., Chen, Y., Dai, W., Yi-Qi, H., Yu-Feng, L., Wei-Wei, T., Qiang, Y., Yang, Y.: Taking human out of learning applications: a survey on automated machine learning. arXiv preprint arXiv:1810.13306 (2018)
2. Zoph, B., Quoc, V.L.: Neural architecture search with reinforcement learning. arXiv preprint arXiv:1611.01578 (2016)
3. Pham, H., Guan, M., Zoph, M., Le, Q., Dean, J.: Efficient neural architecture search via parameters sharing. In: International Conference on Machine Learning, pp. 4095–4104. PMLR (2018)
4. Snoek, J., Larochelle, H., Adams, R.P.: Practical Bayesian optimization of machine learning algorithms. In: Advances in Neural Information Processing Systems, vol. 25 (2012)
5. He, X., Zhao, K., Chu, X.: AutoML: a survey of the state-of-the-art. Knowl.-Based Syst. **212**, 106622 (2021)

6. Waring, J., Lindvall, C., Umeton, R.: Automated machine learning: review of the state-of-the-art and opportunities for healthcare. Artif. Intell. Med. **104**, 101822 (2020)
7. Komer, B., Bergstra, J., and Eliasmith, C.: Hyperopt-sklearn: automatic hyperparameter configuration for scikit-learn. In: ICML Workshop on AutoML, vol. 9. Citeseer, Austin (2014)
8. Google AutoML homepage. https://cloud.google.com/automl/. Accessed 14 Feb 2022
9. Ivakhnenko, A.G., Ivakhnenko, A.G., Lapa, V.G., Lapa, V.G.: Cybernetics and Forecasting Techniques, vol. 8. American Elsevier Publishing Company (1967)
10. Huang, G.B., Zhu, Q.Y., Siew, C.K.: Extreme learning machine: a new learning scheme of feedforward neural networks. In: 2004 IEEE International Joint Conference on Neural Networks (IEEE Cat. No. 04CH37541), pp. 985–990 (2004)
11. Cambria, E., et al.: Extreme learning machines [trends and controversies]. IEEE Intell. Syst. **28**(6), 30–59 (2013)
12. Miche, Y., Heeswijk, M., Bas, P., Simula, O., Lendasse, A.: TROP-ELM: a double-regularized ELM using LARS and Tikhonov regularization. Neurocomputing **74**(16), 2413–2421 (2011)
13. Lendasse, A., Man, V.C., Miche, Y., Huang, G.B.: Advances in extreme learning machines (ELM2014). Neurocomputing **174**, 1–3 (2016)
14. Akusok, A., Miche, Y., Hegedus, J., Nian, R., Lendasse, A.: A two-stage methodology using K-NN and false-positive minimizing ELM for nominal data classification. Cogn. Comput. **6**(3), 432–445 (2014)
15. Lauren, P., Qu, G., Yang, J., Watta, P., Huang, G.B., Lendasse, A.: Generating word embeddings from an extreme learning machine for sentiment analysis and sequence labeling tasks. Cogn. Comput. **10**(4), 625–638 (2018)
16. Miche, Y., Sorjamaa, A., Bas, P., Simula, O., Jutten, C., Lendasse, A.: OP-ELM: optimally pruned extreme learning machine. IEEE Trans. Neural Netw. **21**(1), 158–162 (2009)
17. Khan, K., Ratner, E., Ludwig, R., Lendasse, A.: feature bagging and extreme learning machines: machine learning with severe memory constraints. In: 2020 International Joint Conference on Neural Networks (IJCNN), pp. 1–7 (2020)
18. Liu, N., Wang, H.: Ensemble based extreme learning machine. Signal Process. Lett. 754–757 (2010). IEEE
19. Van Heeswijk, M., Miche, Y., Lendasse, A.: GPU-accelerated and parallelized ELM ensembles for large-scale regression. Neurocomputing **74**(16), 2430–2437 (2011)
20. Grigorievskiy, A., Miche, Y., Ventela, A.M., Severin, E., Lendasse, A.: Long-term time series prediction using OP-ELM. Neural Netw. **51**, 50–56 (2014)
21. Yu, Q., Van Heeswijk, M., Miche, Y., Nian, R., Severin, E., Lendasse, A.: Ensemble delta test-extreme learning machine (DT-ELM) for regression. Neurocomputing **129**, 153–158 (2014)
22. Van Heeswijk, M., Miche, Y., Lindh-Knuutila, T., Hilbers, P.A.J., Honkela, T., Oja, R., Lendasse, A.: Adaptive ensemble models of extreme learning machines for time series prediction. In: International Conference on Artificial Neural Networks, pp. 305–314. Springer, Berlin (2009)
23. Song, Y., et al.: Gaussian derivative models and ensemble extreme learning machine for texture image classification. Neurocomputing **277**, 53–64 (2018)
24. Yu, Q., Miche, Y., Severin, E., Lendasse, A.: Bankruptcy prediction using extreme learning machine and financial expertise. Neurocomputing **128**, 296–302 (2014)
25. Han, B., Miche, Y., Severin, E., Lendasse, A.: LARSEN-ELM: selective ensemble of extreme learning machines using LARS for blended data. Neurocomputing **149**, 285–294 (2015)
26. Lan, Y., Yeng, C.S., Huang, G.B.: Ensemble of online sequential extreme learning machine. Neurocomputing **72**(13–15), 3391–3395 (2009)
27. Dua, D., Graff, C.: UCI machine learning repository. http://archive.ics.uci.edu/ml. University of California, School of Information and Computer Science, Irvine (2019)
28. Fletcher, S., Islam, Z.: Comparing sets of patterns with the Jaccard index. Australas. J. Inf. Syst. **22** (2018)

29. Anguita, D., Ghio, A., Oneto, L., Parra Perez, X., Reyes Ortiz, J.L.: A public domain dataset for human activity recognition using smartphones. In: Proceedings of the 21th International European Symposium on Artificial Neural Networks, Computational Intelligence and Machine Learning, pp. 437–442 (2013)
30. Sakar, C.O., et al.: A comparative analysis of speech signal processing algorithms for Parkinson's disease classification and the use of the tunable Q-factor wavelet transform. Appl. Soft Comput. **74**, 255–263 (2019)
31. Ballabio, D., Grisoni, F., Consonni, V., Todeschini, R.: Integrated QSAR models to predict acute oral systemic toxicity. Mol. Inf. **38** (2019)

Measuring Research Productivity at Scale

Peggy Lindner(✉)

Information and Logistics Technology Department, University of Houston, Houston, TX, USA
plindner@uh.edu

Abstract. The increasing availability of digital data on scholarly inputs and outputs – from research funding, productivity, and collaboration to paper citations and scientist mobility – offers unprecedented opportunities to explore the structure and evolution of science. It has driven the development of a new field - the science of science (SciSci) - which in essence is a transdisciplinary approach that uses large data sets to study the mechanisms underlying the doing of science-from the choice of a research problem to career trajectories and progress within a field. Like any other discipline, the science of science field is dependent to a great degree on their instruments of measurement. The more accurate the instrument, the better researchers will be able to observe specific phenomena. In this paper, we present a framework for data collection and analysis of the large publicly available data in Google Scholar (GS) which has gained ground as a free scholarly literature retrieval source. We will discuss the design and implementation of an automatic web-scraper for GS and its ability to produce reliable scholarly productivity metrics to serve as measures in a large-scale study addressing research questions surrounding the academic promotion & tenure process.

Keywords: Science of Science · Scholarly productivity · Crowdsourcing

1 Introduction

With the recent emergence of an encompassing discipline to study the inner workings of science across disciplines, science of science [11] addresses a variety of questions that can be organized into two main categories:

1. social aspects of modern science
2. cognitive (intellectual) aspects of modern science.

The first topics considers questions such as *What is scientific collaboration?*, *How do research teams form and evolve?*, *How is science organized socially?*, *What is a scientific community?*, *Do various groups of scientists (age, gender, etc.) perform science differently?*, *What is the relationship between success and social practices?* and *What are the career trajectories of scientists?*

K.-M. Björk (Ed.): ELM 2021, PALO 16, pp. 164–169, 2023.
https://doi.org/10.1007/978-3-031-21678-7_16

For the cognitive aspects, a slightly different set of questions arises: *How do new concepts and topics emerge and spread?, How do new research areas/disciplines form?, What is interdisciplinary? and What is multidisciplinary?*

Besides those fundamental questions, Science of Science also tries to address practical concepts like *How should science be founded?* A lot of the questions have been driven by the fact that Science has experienced dramatic changes in knowledge production which is exemplified by an increased pressure on productivity and a shift towards "team science" and interdisciplinary [5].

One of the key factors in the science of science field is the availability of large data sets with information content pertaining to a wide range of variables from scholarly productivity and innovation to composition of collaborative research endeavours. The overall goal is to unveil reproducible patterns that govern individual scientific careers and the workings of science.

In a couple of recent publications, big data resources and methods have been deployed to answer some of the questions listed above. For instance Milojević [7] tried to measure the cognitive extend of scientific literature by parsing titles into phrases and counting the unique phrases in unit quotas of 3000 articles. They found, that 3000 articles which contain fewer unique phrases (X) have a smaller cognitive content than 3000 articles with more unique phrases (Y). The large measurement quote was needed to reduce the stochasticity of the titles. In another article by the same group of researchers [8] which utilized publications in the fields of astronomy, ecology and robotics, they found that the fraction of authors who never become lead authors has risen dramatically. The time over which half of the cohort has left the field has shortened from 35 years in the 1960s s to only 5 years in the 2010s.s. In addition, they found a rapid rise (from 25 to 60% since the 1960s)s) of a group of scientists who spend their entire career only as supporting authors without having led a publication.

While a growing body of literature is tackling science of science research questions, lingering doubts about the validity and utility of the large data sets used in analysis [1] have not been fully addressed. The salient question arises as to which service is the most reliable or useful, however the foundational decisions and operational principles of these services are often proprietary or difficult to cross-compare. The problems become even more pronounced when topics around scholarly productivity are the subject of the research. The potential for the typical metrics of impact to not equally or equitably quantify work across disciplines is huge and will also depend heavily on the chosen data base. To take an example, even a simple statistic such as the H-Index [4] or number of publications authored by an individuals varies significantly across service providers (see Fig. 1).

Nevertheless, the availability of Goggle Scholar as a free publication data base makes it a compelling choice for many researchers [2,3]. As a free resource, Google Scholar doesn't provide an API to collect the data in an organized fashion and this lack of accessibility has prevented researchers with non-technical skills from widespread utilization of large scale Google Scholar data. In the following, we will present the design and implementation for a framework to collect and

B-9565-2008/0000-0002-5219-1625 (https://publons.com/researcher/822892/jodi-lindsay/)

	2013	2015	2017	2019*
Web of Science	25	28	34	41
Google Scholar	28	33	42	50

(of these, 44 were published before 2013)

Fig. 1. Example illustrating variation in raw publication counts across two different data bases – Web of Science and Google Scholar.

organize Google Scholar data with an author-centric view and discuss it's utility in a first case study.

2 Collecting Google Scholar Profile Data at Scale

Google Scholar allows to search all scholarly literature from one place, explore related works, citations, authors, and publications. Most academics however know Google Scholar through the Google Profile view which consolidates publications attributed to a person. The creation of the profile itself has to be initiated by a person and that person becomes the "owner" of a profile. The attribution typically happens automatically through a name matching in the list of authors for a publication. Profile owners can also manually add publications and list their affiliation and research areas as part of the profile. Lately, a list of co-authors can be added to the information presented on the profile page. In addition, a Google profile page lists a couple of metrices such as H-index, the i10 index (index based on the number of publications with at least 10 citations) as well as total citation counts (current and over time). The profile page and the derived metrics are based on the publication record attributed to a scholar. That publication record is collected automatically through Google's vast content scanning mechanisms. A curation/cleaning of the individual publication records can only be done by profile owners themselves.

Unlike the keyword/topic based search for a list of publications, starting out with a list of scholars and collecting profile information allows for a more author-centric organization of the data which is often required when studying questions related to productivity/impact of individuals. Therefore, we have designed a framework that requires seed information in the form of Google Scholar Profile URLs (see Fig. 1). We use the initial list of URLs to fill a queue that feeds individual scraper instances to initially fetch and store information listed on a GS Profile page. The retrieved list of publications will then feed a second queue to fetch and store a detailed publication record inclusive the citation history for that record. Every fetch creates a new read request and we have opted to deploy a slow but steady process for the data collection which allows to collect around 200 profiles/complete publication records for one scarper instance per week. Each scraper operates independently by picking up jobs from the seed list imported into the DB allowing the scraper farm to grow the data over time.

The framework has been implemented using R [10]. We have expanded an original idea presented in [12] to allow for the distribution of scraping workloads and therefore afford the feasibility of large scale GS Profile data collection. Our implementation stores the data directly into a cloud hosted database to support maintenance and expansion of the data collection over time. We plan to scrape records once every year and hence will be able to update and keep a history of the data. The implementation is capable to follow Google Profile links that have moved to a new URL. We also keep a record of retracted/removed Google Profile pages.

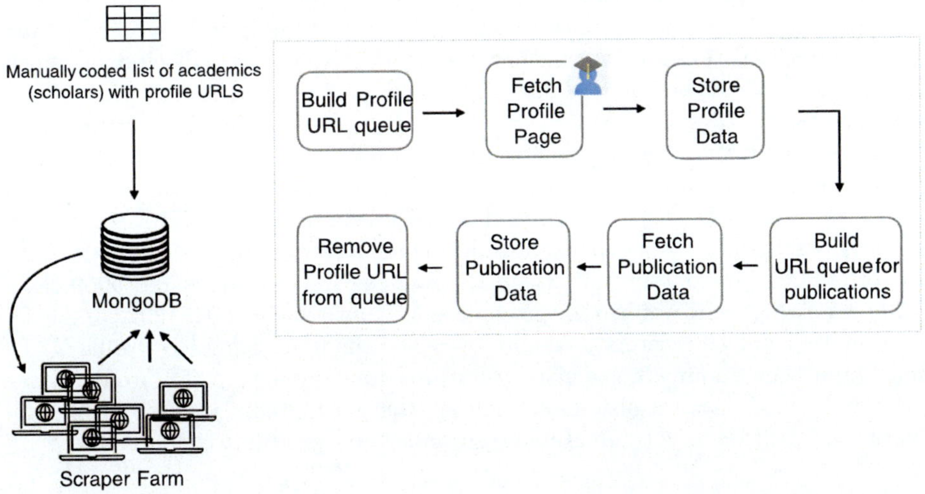

Fig. 2. Schematic process for the Google Scholar scraper.

The scraped profile data can be used as is but cleaning/pruning of the publication records is a must. We are currently deploying a rules based implementation that uses publication record key variables depicting the type of publication in conjunction with filtering other fields like title and description. Additional cleaning can tackle issues with "time bending" (citation records exist before the actual publication date) in GS citation histories.

3 Case Study

With the scraping farm, we initially collected 5500 profiles and publication records which yielded roughly 1 Mill publication records. Using that data organized by scholar, we have implemented scripts to calculate H-index over time based on historical citation counts. We can also extract the number of publications in top publications for certain fields. Filtering is based on lists and currently we can provide data for Business and Computer Science. The scripts are an integral part of the overall framework and can be utilized at any point of data collection or configured to be used with subsets, e.g. by discipline. In addition to extraction of baseline productivity data, the data structure allows for the creation of co-author networks.

A recently founded NSF project [9] is investigating critical questions around ERLs that have been neither theoretically nor empirically examined. External review letters (ERL) provided by arm's length reviewers are critical components of promotion and tenure (P&T) processes. Described as a vital factor determining the fate of a researcher's career, academic administrators often view ERLs as the most critical components of P&T portfolios, providing qualitative, independent evaluations of candidates' past accomplishments, reputation, and the prospect of sustained contributions. As part of the study, we are trying to develop a holistic approach to measure productivity across disciplines which doesn't mean the creation of a new single new metric but the provision of different measures that can lead to comparable result for scholarly productivity. We have started the scraping process and in addition to productivity measures, we plan to extract authorship hierarchy information (e.g. first author, second author etc.). Another aspect will be the integration of scalable NLP techniques like word embedding and sentiment analysis for letters and/or large scale analysis of publication content [6]. The final data set will contain around 9000 cases (3000 candidates who applied for promotion & tenure during 2017–2022 and 6000 external review letterwriters). We have received the first coded GS Profile URLs and found that about 60% of scholars in the data set have a GS profile. Since this hand coded data set also allows us to search for matching author profiles on Scopus, we will be able to create a cross-walk between the two databases.

4 Discussion

The availability of large scale publication data bases has spurred research in Science and Science and led to a new understanding of the underpinning of scientific discovery. We presented a framework to collect Google Scholar data at scale and organize the data to be useful for analysis in multiple scenarios. When using this data as measure for scholarly productivity, future work will have to constantly address questions with respect to validity in the ever evolving expansion of the available data bases. First internal consistency will have to be evaluated with respect to variation in results, using GS data to calculate H-index versus the h-index displayed on a Google Scholar Profile page. The more pressing issue however, will be to address external in-consistency across data services (WoS, Scopus, Microsoft Academic, etc.) which is caused by different coverage e.g. by field and what record are included. Our preliminary work has shown some evidence that "correlation" and "ranking" are reasonably stable across platforms despite variance in the absolute numerical value of metric.

References

1. Gusenbauer, M.: Google Scholar to overshadow them all? Comparing the sizes of 12 academic search engines and bibliographic databases. Scientometrics **118**(1), 177–214 (2018). https://doi.org/10.1007/s11192-018-2958-5
2. Halevi, G., Moed, H., Bar-Ilan, J.: Suitability of Google Scholar as a source of scientific information and as a source of data for scientific evaluation-Review of the Literature. J. Inf. **11**(3), 823–834 (2017). https://doi.org/10.1016/j.joi.2017.06.005, https://linkinghub.elsevier.com/retrieve/pii/S1751157717300676
3. Harzing, A.W., Alakangas, S.: Google Scholar, Scopus and the Web of science: a longitudinal and cross-disciplinary comparison. Scientometrics **106**(2), 787–804 (2016). https://doi.org/10.1007/s11192-015-1798-9, http://link.springer.com/10.1007/s11192-015-1798-9
4. Hirsch, J.E.: An index to quantify an individual's scientific research output. Proc. National Acad. Sci. **102**(46), 16569–16572 (2005). https://doi.org/10.1073/pnas.0507655102, https://www.pnas.org/content/102/46/16569
5. Jamison, A.: Knowledge making in transition: On the changing contexts of science and technology. In: Nordmann, A., Radder, H., Schiemann, G. (eds.) Science Transformed?: Debating Claims of an Epochal Break, chap. 8, pp. 93–105. University of Pittsburgh Press (2011)
6. Lauren, P., Qu, G., Yang, J., Watta, P., Huang, G.-B., Lendasse, A.: Generating word embeddings from an extreme learning machine for sentiment analysis and sequence labeling tasks. Cogn. Comput. **10**(4), 625–638 (2018). https://doi.org/10.1007/s12559-018-9548-y
7. Milojević, S.: Quantifying the cognitive extent of science. J. Inf. **9**(4), 962–973 (2015). https://doi.org/10.1016/j.joi.2015.10.005, https://www.sciencedirect.com/science/article/pii/S175115771530081X
8. Milojević, S., Radicchi, F., Walsh, J.P.: Changing demographics of scientific careers: the rise of the temporary workforce. Proc. National Acad. Sci. **115**(50), 12616 (2018). https://doi.org/10.1073/pnas.1800478115, http://www.pnas.org/content/115/50/12616.abstract
9. National Science Foundation: Award abstract # 2100034. https://www.nsf.gov/awardsearch/showAward?AWD_ID=2100034 (2021), accessed = 2021-11-30
10. R Core Team: R: A Language and Environment for Statistical Computing. R Foundation for Statistical Computing, Vienna, Austria (2021)
11. Wang, D., Barabási, A.L.: The Science of Science. Cambridge University Press (2021). https://doi.org/10.1017/9781108610834
12. Yu, G., Keirstead, J., Jefferis, G., Getzinger, G., Cimentada, J., Czapanskiy, M., Makowski, D.: Analyse Citation Data from Google Scholars (2021), r package version 0.2.2

Author Index

K.-M. Björk (Ed.): ELM 2021, PALO 16, pp. 171–172, 2023.
https://doi.org/10.1007/978-3-031-21678-7